Physics Education

Physics Education

R S REDDY

COMMONWEALTH PUBLISHERS
in association with
DR. ZAKIR HUSSAIN INSTITUTE FOR NON-FORMAL AND CONTINUING EDUCATION

Physics Education

First Published : 2012

ISBN 81-7169-473-X

Published by :

Ajay Verma
For COMMONWEALTH PUBLISHERS
4831/24, Prahlad Street,
Ansari Road, Darya Ganj,
New Delhi - 110002
Ph. : 23272541, 23257835
Fax : 91- 011- 23257835
e-mail : campusbooks@hotmail.com

Typesetting by :

Arjun Computers
Delhi - 110051

Printed at :
Efficient Offset Printers
New Delhi

Preface

This book has been created to provide access to information about contemporary education. Practitioners and students at all levels in education have a need to know what is happening today.

The book will be of use to graduate and post graduate students in education and to practicing teachers, administrators, librarians and planners. The chapters and the further sources of information cited in this book should lead the reader to thousands of people and documents for either research or programme planning purposes.

This book is addressed to administrators, planners and educators working in the field of education and training with a view to stimulating interest and attention in the areas of education and its related fields. It is also addressed to a growing number of teachers and instructors who will be practitioners in education and who will need to be acquainted with the modern aspects of educational practice and development. Many ideas, generalisations and discussions presented in this book should also prove useful to employing organisations committed to provide training facilities within their establishments—leading to effective mutual participation by institutions and organisations.

The editor wishes to thank the contributors, as well as those organizations that gave permission to publish their extracts, chapters etc.

Editor

Contents

1

Differences in Students' Perceptions of Learning Physics

Recent research on student learning in higher education has highlighted the effect that students' perceptions of the nature of learning and understanding of their discipline has on their consequent understanding of the subject matter. This present project grew out of a programme aimed to help first-year students in physics develop their ideas on the nature of physics and how to study it. One of the issues to emerge from that project was the need to develop a more detailed understanding of students' perceptions, to be used both in further planning of the program and in its evaluation.

This paper reports on the results of that study. It involved a before (340 students) and after (110) survey of first-year physics students at the University of Sydney. They were asked to complete an open-ended response survey, including questions on how they go about studying physics and what they thought the study of physics was about. A phenomenographic methodology was used to identify categories of description for each of the questions.

The paper will describe the categories in some detail and the distribution of responses across the categories. It will also discuss a number of issues to emerge from the study and some implications for the practice of teaching physics.

Recent research on student learning in higher education has highlighted the importance that *(a)* students' perceptions of the nature of learning and understanding of the discipline, and *(b)* their approaches to the study of the discipline, have on the quality of student learning outcomes, and in particular, qualitative learning outcomes.

For example, Saljo identified five qualitatively different, and hierarchically related, concepts of learning for a group of adults. The concepts were:

1. a quantitative increase in knowledge;
2. memorizing;
3. the acquisition of facts, methods, etc. which can be retained and used when necessary;
4. the abstraction of meaning;
5. an interpretive process aimed at understanding reality.

The former concepts are based upon a view of learning emphasizing reproduction while the latter concepts emphasise understanding. Earlier, Marton and Saljo had identified two qualitatively different approaches to reading a text-a surface approach and a deep approach, and showed that these approaches are closely related to the outcomes of reading the text, van Rossum and Schenk showed that the conceptions of learning identified by Saljo were closely related to the approaches to learning identified by Marton and Saljo.

At about the same time, using more quantitative

methodologies, Entwistle and Ramsden showed that students' perceptions of the teaching and learning context were closely related to the way students' approach their learning in that context. Thus, the research has shown a chain of relationships linking concepts of learning, perceptions of teaching and learning contexts, approached to learning and quality of learning outcomes. These links have been verified by latter research using both qualitative and quantitative research methodologies.

Much of the research from this perspective in physics has looked closely at the relationship between approach and outcome, but has not looked at students' broader perceptions of physics. The importance of students concepts, perceptions and approaches to the quality of their learning outcomes is well established. The present research addresses the important question of how they are constituted within the particular discipline of physics, and how are they distributed within a particular course in physics.

Specifically, this present project grew out of an earlier programme developed by one of the authors aimed at helping first-year university students in physics develop their ideas on the nature of physics and how to study it. One of the issues to emerge from that project was the need to develop a more detailed understanding of students' perceptions, to be used both in further planning of the programme and in its evaluation. Thus the specific questions we are addressing in this study of student learning in first-year physics courses are:

What are student perceptions of the nature of learning physics?

How do these perceptions change over their first year of study?

How are their perceptions related to their learning outcomes?

In this chapter we report on the results of the first of these questions.

Methodology

The results in this paper are part of a larger study conducted in first-year physics at a large, well-established, metropolitan university. In that study, all first-year students enrolled in the first year of the physics major stream were surveyed at the beginning of their first semester of study and again at the end of their second semester. The first survey was designed to identify first-year students' prior perceptions and approached to studying physics, while the second survey focused on their experiences over the previous year.

The sample consisted of 330 responses to a survey conducted in week three of the students' first semester of study in the University.

The survey consisted of six, half-page open written response questions. The questions were designed to elicit a personal response from the students. The questions were phrased in words which students could easily relate to and which would elicit responses in their own words, reflecting their experiences of studying physics. This paper is based upon responses to four of these questions. They were:

1. Why do you think some people seem to learn physics more easily than other people?
2. How do you go about learning physics? Say what you actually do rather than what you think you should do.
3. If you had a friend who had never done physics

before and they asked you to tell them what the study of physics involves, what would you say?

4. If someone asked you for advice about how to learn physics, what advice would you give them?

In analysing the responses, the focus was first of all to identify as set of categories of description for each question and then to classify the responses in terms of these categories of description. The analysis was based upon a phenomenographic methodology, and involved the three authors identifying categories for a small subset individually; meeting, discussing and revising the categories, piloting the categories on a second subset, meeting, discussing and revising the categories. This procedure continued until a reasonably stable set of categories was produced..Once the final descriptions were agreed upon, one last sample of 25 responses was selected and categorized by two of the researchers and a research assistant. Agreement on the categorization on all four questions ranged from 72% to 92%, which is accepted as a reasonable level of agreement for such research. A research assistant then classified the remaining responses according to these categories.

Results

The categories of description and distribution of response are shown for each question in turn.

Question 1

The categories of description of the analysis of the first question-why do you think some people learn physics more easily than other people?-were as follows (with examples):

Category 1 - responses based upon innate ability or hard work (*n* = 165, 55%), e.g.

> 'They are smarter'.
>
> 'Different people have different capacities for different activities, some more 'suited' to physics'

Category 2 - response based upon person seeking understanding or relating to other subjects and/or previous experience (n = 68, 23%), e.g.

> Some people try to understand the concepts involved rather than just memorize equations'.
>
> 'Some people can understand things more easily and are able to relate new ideas to Previousones. Others find that they cannot understand the fundamentals and give up, therefore not being able to understand more complicated ideas'.

Category 3 - response based upon relating to real world experiences, or of building models about the real world (n = 10, 3%), e.g.

> 'I think it has something to do with the fact that some people can find examples of physics or physics applications in their everyday life. Also I find it helps when teachers or lectures use problems or examples that the student can relate to'.
>
> 'People who apply physics to everyday situations seem to understand it and learn it more easily'.

Category 4 -response based upon interest in the subject (*n* = 56,19%), e.g.

> 'People learn easier if they have a genuine interest

and do not feel forced to study it'.

'They are more interested'.

From the phenomenographic perspective, categories 1, 2 and 3 are logically related through inclusiveness. That is, for example, the seeking of understanding (category 2) requires some innate ability and some hard work (category 1). But innate ability and hard work are not sufficient, making category 2 inclusive of category 1. Similarly, category 3 is inclusive of categories 1 and 2. Category 4, however, does not fit into this structure.

Its focus is on interest, and could logically be associated with any of the three other categories.

It may be worth noting that of the three logically related categories, category 1 is independent of the subject matter, and in that sense is generic. Categories 2 and 3, however, have more of a subject matter focus. For these categories it is how you relate to the subject matter that is important. The focus is not so much on how much you do, but what you do when you work.

The distribution of responses suggests that the majority of students (55%) enter the physics course believing that hard work and ability are the major reasons for success, not how you study and relate to the subject matter. To do better than others, you need to be brighter and to work harder, not that you work in any way differently.

Question 2

The categories of description of the analysis of the second question-How do you go about learning physics? Say what you actually do rather than what you think you should do? -were as follows (with examples):

Category 1 -explanation based upon attendance and/or reviewing notes and/or learning formulas and/or doing exercises (*n* = 218,75%), e.g.

> 'Doing the assignments, listening in class, studying for exams'.
>
> 'I try to look at examples first. Then try some questions to build confidence, then when I feel confident, I do a lot more questions to be good at it'.

Category 2 - response based upon seeking under-standing-seeing how principles work, discussing with other students (n = 61,21%), e.g.

> 'In class I take notes of the important details without being a slave to them. I try to understand the concepts there and then rather than adopting a 'she'll be right' attitude. Any areas which intrigue me or of which I have incomplete understanding I will ask questions about'.
>
> 'Go to lectures and try to understand the concepts behind what is being taught. This may also require some time at home staring at the ceiling puzzling it out until it 'clicks'. Ones this understanding is in place, any problems/exercises should be relatively straightforward. Any difficulties with problems which are then thought through and understand reinforce the overall understanding'.

Category 3 - response based upon relating to real world experiences, reading around the subject etc. (n = 12,4%), e.g.

> 'Examples, applications to the physical world. To learn physics, better, to understand physics, needs visualization of many concepts so experiments and observations substantiate learning. Many areas just need various examples to explain it more clearly but in a different context. Thus demonstrating the basics. Problem solving is the key'.

> 'I enjoy reading about physics, e.g. in New Scientist, Astronomy magazines and in books. I also try simulating physical problems on computers-I can't really tell the difference in the time I spend studying physics and just having fun'.

Again the categories form a logically inclusive hierarchy, with category 3 being more sophisticated and inclusive of categories 1 and 2. While attending classes, reviewing notes and learning formulas may be necessary, it is not sufficient.

In category 1, the focus is on reproduction, while in categories 2 and 3 the focus is on seeking understanding. The distribution of responses suggest that three-quarters of the entering students approached their studies by seeking to reproduce rather than seeking understanding. In this sense, category 1 represents a surface approach to study, while categories 2 and 3 represent deep approaches.

Question 3

The categories of description of the analysis of the third question-If you had a friend who had never studied physics before and they asked you to tell them what the study of physics involves, what would you say?-were as follows:

Category 1a-response based upon physics being about facts and formulas and/or hard work (n = 37, 12%), e.g.

> 'It involved use of mathematics to work out problems that are present in many areas'
> 'Learning of formulas, a logical mind'.

Category 1b - response based upon physics being about the study of the physical world (n=121,41%), e.g.

> 'The study of the world around us'
>
> 'Investigating how the universe works'.

Category 2 - response based upon physics being about the relationship between mathematics and the physical world and/or understanding the underlying principles governing the behaviour of the physical world(n=91,31%), e.g.

> 'If you're clever it involves coming to understand principles about the physical world. If you're dumb it involves learning lots of rules and doing lots of questions even though you can't understand why they work'.
> 'A lot of hard work, along with understanding of why some things happen'.

Category 3 - response based upon physics being about an integrated creative process of developing models and a language to describe observed behaviour of physical systems in the physical world (n = 12,4%), e.g.

> Try to see where physics fits into life. Doing problems to help me understand how and why'.
>
> 'A process of successive approximation, which attempts to construct an internally consistent, and experimentally consistent explanation of the

phenomena of the natural world. It gives no pretence of delivering a 'true' representation of the universe'.

These categories again form a hierarchical structure through logical inclusiveness, with categories 1a and 1b being relatively unsophisticated views of what the study of physics is about, with categories 2 and 3 having an increasingly sophisticated focus. The distribution of responses again suggests that the majority of students have a relatively unsophisticated view (53% in categories 1a and 1b).

Question 4

The categories of description of the analysis of the fourth question-If someone asked you for advice on how to learn physics, what advice would you give?-were as follows:

Category 1 - response based upon attendance and/or reviewing notes and/or learning formulas and/or doing exercises (n = 175,65), e.g.

> 'Do heaps of problems'.

> 'Learn your formulas and laws. Everything comes from them'.

Category 2 - response based upon seeking understanding-seeing how principles work, discussing with other students (n = 77,295),, e.g.

> 'Read the text for a superficial understanding of the work. Greater understanding comes from doing problems.'

> 'Try to understand what is actually happening rather than just memorizing'.

Category 3 -response based upon relating to real world experiences, reading around the subject etc (n=17,6%), e.g.

> 'Read magazine articles, or rent documentary videos to spark a real interest in physics by having a more realistic understanding of its real-life relevance. Also treat problems as fun puzzles to be solved (like crosswords, chess problems etc)'.

> 'Don't read a book, don't copy huge amounts of notes or mathematical derivations of formulae. Try and think about the physics behind everything that surrounds you'.

The hierarchical structure in the set of categories of description is again evident for question 4. Categories 2 and 3 represent the recommendation of deep approaches to study, while category 1 represents the recommendation of a surface approach. The majority of students recommend a surface approach (65%).

Discussion and conclusions

In discussing the results of this study, three issues seem to stand out. They are: (a) the high proportion of low-level responses; (b) the finding that students say physics is about the study of the physical worlds, yet do not approach their study in this way, or recommend that others do; and (c) the importance of interest in students' attribution of success to other students. Having discussed the issues, some implications for the practice of teaching physics will be outlined.

Issues

The results show that the vast majority of students'

responses were classified in the low-level categories. Most students thought that success in studying physics was about innate ability and/or hard work, rather than how one goes about studying (a view not supported by the research findings detailed in the introduction). Most students adopted a 'surface' approach to learning in terms of attending classes, reviewing notes, learning formulas and doing exercises. Few indicated that they were seeking understanding in terms of how the major principles worked or relating their knowledge to real world experiences-a 'deep' approach. Interestingly, most students saw physics as being about the study of the physical world-an issues dealt with later in this discussion. Finally, in terms of the way they recommended to others how they should approach their studies, students recommended a surface approach.

The second issue is that most students acknowledge that physics is about the study of the physical world, but then only a minority approach their studies in terms of seeking to understand the physical world. The issue is that a deep approach to the study of physics involved students relating the new knowledge to their real world experiences, but very few approach their studies in this way. The explanation may be that they approach their studies in ways to reproduce for assessments rather than trying to understand, or that the 'study of the physical world' is a throw-away line that does not apply to their world but to an abstract idealization.

Finally, it is of interest to note that a substantial number of students attributed the success of others to them having an interest in physics, while interest did not figure in the responses to the other questions. Why students would see interest being important to others but not so important to themselves is not clear. It may, of course, be in the nature of the questions posed. But the issue of affect has not been well dealt with in phenomenographic studies in the past, and is one which would be looked at more closely in the future.

Implications for practice

Changing student's understanding of what physics is about and the way they should approach their studies of it in the sense described in this paper, requires substantial changes to the context in which leaning takes place. Assessment practice, course structures (including the sheer amount of content to be learnt), teaching methods, etc all have an impact on students' understanding and approached to study. More specifically, however, thee are a number of ways of addressing these issues in teaching physics. Two examples, and the principles underlying them, will be outlined here. Both examples have been, and are being, used by the authors in their teaching of large, first-year university, physics classes. The first is an example of relating students' understanding to their experience of the real world, and the second is an example of helping students develop the way they are studying physics.

In lectures of 100 or more students, the amount of time devoted to formal theory development can be reduced by providing students with detailed study guides to their textbook in which the theory is developed. Lecture time can then focus on applications of physics to real world situations. For example, we have used 'buzz groups' in mechanics lectures to encourage students to focus on their real world experiences of the subject matter. In these students are given a problem to consider in their first lecture-that of the forces acting on a car accelerating up a hill-and are asked to discuss the problem with their neighbour for 3 to 5 minutes. The results of the discussion are summarized on the board by asking students at the front, in the middle, and at the back of the lecture theatre to describe their force diagrams. Students quickly become aware that their fellow students explain the forces in ways different to them, and that thinking about how the physics they are studying can be applied to seemingly simple real world problems can expose their misunderstandings of the basic concepts.

Similarly, helping students examine their approaches to leaning involved helping them see that their fellow students study in different ways with different intentions. Students are asked in a lecture or a tutorial to write down what they think the study of physics is about, and to describe how they have approached their studying in the previous week. They compare their responses with those of their neighbour(s), and note differences. These are summarized immediately as in the previous example, or the lecturer or tutor collects the responses, reads through them, and comments on them in the following class. Again, students quickly become aware of the fact that their fellow students study in ways different to them-which surprises them-and the lecture or tutor can comment on the way they see most appropriate.

In both these examples the structure of the activity is to encourage students to engage in a task based upon their own experience, to compare their responses to those of their fellow students, and receive feedback and advice from their lecture or tutor. Just telling students what physics is about and how they should approach their studies of its is rarely successful. Students need to confront their prior understandings, and comparing their understandings to those of other students is a very powerful means of achieving this.

2

Multiplying Physics Enrollment– Strategies that work

Articles have appeared recently in the Physics Teacher concerning the more nontechnical aspects of teaching physics and the prospect of increasing physics enrollment, especially among groups of students who would not normally take a course in physics. We have spent the past five years creating and advertising a physics program that appeals to the majority of students, but sacrifices neither rigor nor content. We have a two-pronged approach: first to extensively publicize the class at school and second to more the course our of my classroom and into the home.

Advertising Physics at school

One strategy we use is to never do a lab inside when it can be done outside. Seeing the physics students working and yet totally enjoying their time outside is a great advertisement to the younger nonphysics students who gaze forlornly out their classroom windows at us. One example of an outside activity we do is the Rocket Lab in which students build and lunch model rockets. When we introduce vectors. We have students find the displacement between two cones located on the school track. They do this by starting at one cone and walking in the opposite direction around the track to ward the other cone, counting their paces in straight-line segments and measuring their directions with compasses. We often take students out for a quick demonstration that lasts only ten minutes, but even in these short venture they seem more energetic when they return, and the inevitable observers in the windows are left with a favorable impression of what goes on in physics.

For the last several years we've had my students design physics buttons to be worn both on and off campus. Scattered throughout this note are illustration s of buttons from the last four years. Students love being part of this. The buttons announce in an indirect way that the bearer is a Tamalpais student of physics-no small statement for the student who has had academic difficulties in the past. The excitement of wearing the buttons fades a bit after a few weeks and they normally end up getting placed on backpacks. But we encourage students to all wear their buttons on certain special days that we call "physics Solidarity Days." It's a remarkable experience to walk about campus on these days. Everywhere you go you see button-bearing students; some alone, some together with other physics students, some amongst groups of athletes or thespians. The message to the observer is clear. There are lot of physics students on this campus, they come from every group and clique, and there is something special about taking that class.

Another idea we use for advertising my program at school is with physics bulletin questions. Once or twice a week. I'll put a short and semicryptic physics statement or question in the daily bulletin. For example:

You've probably always believed your car had one accelerator. 120 Tam physics students can explain why each ear actually has three accelerators. Ask one!

The Earth has a mass of 5.96 X 10 24 kg, But most physics types would say it weighs less than 200 pounds. Ask any of Tam's 120 physics students to explain. Twinkle, twinkle, little star, how I wonder.. why you twinkle. Wonder no more. Ask a physics student.

The bulletin inserts have a nice effect. A teacher reading the bulletin may ask if there is a physics student present to answer the query. If so, the student has the opportunity to shine as he or she explains the question or statement. Even if it's a class with no physics students. The bulletin plants a seed of interest. The students hear about a class that answers relevant questions. So the message is repeated in one more venue-don't miss this course in physics; don't be left out.

Advertising physics at Home

The second part of my program is to acquaint parents with the course in physics at Tamalpais High. This program has two components; Dinner Table Physics and Dinner with the Family.

Once or twice a week I offer my students an extra-credit opportunity. The extra credit is worth 10% of a lab and 1 call it Dinner Table Physics. I ask students to present a specific physics demonstration at home. It might be a replica of one of my Newton's first Law demonstrations or the formation of a real image from a shaving mirror or the sound wave resonance from a wine glass. The requirement is that students perform the demonstration and teach the physical phenomenon illustrated. Then, to receive the extra credit, the parents must write about what they've learned and the student must turn it in the next day. I deadpan with the students that my objective is to draw them closer to their parents in this sometimes difficult adolescent period where parents (and their children) are often tolerated more than appreciated. But, two real benefits are that students learn the material better and parents hear what is going on at school in physics. Teachers understand better than any others how well a concept is learned if it has to be taught. My students relate that they really "get it" after they have had to explain a concept to their parents in a few different way before they understand.

The other component of my advertising-physics-at-home program is Dinner with the Family. It has been a wonderful experience. Four years ago I decided that I would visit a different family's home each Wednesday night. I announced in each of my classes that I was available to come over for dinner and that they could put their names on a sign-up list on one of my chalkboards. Finally, after a couple of weeks and no takers. I coerced the president of the student body (a student of mine at the time) to invite me. Remarkably, the sign-up list filled the next day and it has been full ever-since. I now have 30 to 35 dinners a year with families of students. Sometimes two or more students will get their respective families together for a big multifamily dinner. Some dinners are modest. Others are elaborate and formal. Still others celebrate the ethnic heritage of the family. But no matter how many or how few people are present, no matter the type of home or the style of food served, parents are amazingly appreciative that I am dining with them. All the families want to hear the story of how I became a physics teacher and everyone wants to talk about physics. By the end of the evening parents and siblings are pretty excited about physics. World gets around-the parent underground is amazingly efficient and interconnected. The net effect is that the physics program is advertised very positively.

As anyone who teaches physics knows, physics is a whole lot of fun. The problem has always been to convince the masses that physics is not only fun, bout comprehensible. We have a student population of 900 in grades 9 through 12. Using the strategies outlined in this paper, the typical enrollment in our physics program has increased over five years from approximately 30 students per year to 120 to 150 per year.

3

The Human Laser

Here is a participation activity for students of all ages that simulates the processes that occur in the buildup of a laser pulse. Through playacting, students are exposed to the concepts of coherence, monochromaticity, stimulated emission, pulse buildup from background fluctuations, and the properties that dictate the shape of a typical laser pulse.

I developed this "analogy activity" when asked to speak to a group of young people (ranging from first-to tenthgraders) about some of my research. After a brief show-and-tell session in which the students had an opportunity to visualize the structure of a laser pulse on an oscilloscope, I got them involved in producing a human-size model of a laser. Of course many in the audience were not familiar with the concepts of electromagnetic radiation and atomic structure, so he model we produced was simplistic. But it served to explain the fundamental concepts of laser amplification and the properties of radiation. It's an idea other physics teachers may find useful.

Before your students go 'on stage', take a moment to

describe the fundamental of lasers, which can be found in most general physics tests. To help your students gain perspective, make a blackboard sketch of the device they are about to simulate-a solid-state, optically pumped laser such as a flashlamp-pumped Ruby laser (the first working laser system). I'm assuming that you will have about 30 students involved.

Ask for volunteers for the following roles: the end mirror, the output coupler, the background fluctuation photon (BF photon), and the optical pump. It's interesting to watch the students struggle with the concepts of each role, perhaps having just heard the terms discussed for the first time, and trying to decide which role is right for them. After these 'leads' are chosen, the rest of the group is assigned to act as the laser ions. Now move your class into a long hallway if you can.

Since the model is of a solid-state laser, arrange the "actors" into two straight lines and make clear that they are not to move their feet since they are now ions in the lattice structure of the crystal. Next, the students playing the roles of end mirror and output coupler are arranged to produce the laser cavity. They are positioned approximately ten feet from each end of the student ion lines. The 100% reflective student end mirror is instructed to send any student coming that way back through the group; the 75% reflective student output coupler is told to let one out of every four students pass out of the cavity. The student who volunteers to be the optical pump will control the room's light switch. The exercise begins with the lights off. The optical-pump student will turn on the lights on your command. The BF photon is to march into the crystal chanting the school's colors ("green, green, green,..") when instructed. If you have a small class, one student could play both the optical pump and the BF photon.

With the classroom lights off, the students representing the ions are to act in a relaxed manner. When the lights are turned on, they are to become and remain excited. This can be done by standing in their places and waving their hands in the air while proclaiming loyalty to the school's predominant color. Explain that this simulates the optical pump and the reaction of the ions to move into a long-lived excited state. The group must at this point agree on the initial single pass gain of the laser medium. Since this parameter depends on the stimulated emission cross section of the laser irons and the density of these irons in the crystal, it varies, and thus must be decided upon before continuing. A reasonable value for this simulation is two, three, or four, depending on the size of the class. For this description a gain of two is assumed. Make it clear that when a student photon taps a student laser ion, the ion must stop its excited behavior, relax, and follow the photon, marching instep and chanting the same color. At his point the student laser ion is changing roles from that of an excited ion to one of the photons marching back and forth through the laser cavity. Mention that this is analogous to the process of Light Amplification by Stimulated Emission of Radiation, which just happens to be the meaning of the acronym LASER. Tell your "ions" to march slowly so they can see the evolution of the pulse; even after they leave the cavity they should continue to march slowly and stay with the group with which they existed the cavity. Inform them that you will over-see the passes through the crystal, as a sort of "physics police officer," so that they do not break any of the laws of physics.

Everyone is now ready to simulate the buildup of the laser pulse. The lights are turned off and the students take their places, calm and quiet. When the lights are turned on, they begin to wave their hands in the air and shout "I like green." The BF photon enters the cavity and begins the pulse build-up process. One student is tapped and follows the BF photon out of the crystal towards the end mirror, both marching in step and chanting "green, green, green.." At the end mirror both students are sent back into the crystal and each chooses another student to join in the march. At the output coupler one student is allowed to pass and the other three enter back into the crystal. After another double pass, twelve students approach the output coupler and three are allowed to leave the cavity. The remaining nine students go back towards the crystal. In this simulation there will not be enough students for each of them to have a partner. This is your chance to point out that this is analogous to the depletion of the laser gain as the pulse builds in the cavity. They "stimulate the emission" of the remaining students and reflect back and forth through the cavity until all the students have escaped. The output coupler is instructed to let a single person escape at each of the round-trips, as long as the student photon has made at least four round-trips. This gives the students some idea of the term "photon lifetime". After the laser cavity is depleted of students, the group continues marching and chanting. They should be arranged in lines along with the students who escaped the cavity with them. Now ask the group to look around them and notice that they have formed a pulse-shaped group.

Keep the students in this arrangement while you draw the analogy: the properties of laser radiation such as coherence and monochromaticity are depicted by the marching in step and the chanting of a single color, respectively. Descriptions of the more difficult concepts of stimulated emission, pulse build-up, the sweeping out of gain, and the trailing edge of a laser pulse are also facilitated by observing their arrangement. In addition, the exercise illuminated the relationship of the pulse length to the length of the cavity and the gain of the laser medium. By increasing the length of the cavity, the round-trip time (t_{rt}) for a photon increases, causing an increase of space between the student groups and an overall stretching of the laser pulse. With an increase in the gain of the laser medium, the excited ions are swept out in fewer round-trips, which results in a higher peak and shorter pulse. To help see this effect, you may want to record the arrangement on the board and then change the gain of your system to produce a new pulse.

Since this exercise was first used with the Young Astronauts Program at the Glenelg County School of Maryland, I have employed the analogy successfully with students of all ages. It has been adapted to explain the operation of several different types of lasers, including the common constant wave helium neon laser. It is a great way to get students up out of their chairs and involved in the class.

4

Laser Speckle Experiments for Students

The Speckle produced when a laser beam is scattered from a rough surface can be used in undergraduate experiments to demonstrate various properties of light and also to test for ulty eyesight.

The grainy appearance of a scattering surface illuminated by coherent light is a phenomenon readily observed by anyone using lasers. The understanding of how this laser speckle is generated and how it can be used for a variety of purposes arms the basis of a series of undergraduate experiments that have been devised in Physics department of the University of Survey. With the aid little more than a 1 mW helium-neon gas laser and readily available student optical equipment the experiments properties of waves, polarization effects, the focusing of Gaussian beams, Brownian motion, and, of particular interest to medical physics students, the detection and measurement of accusing defects in the human eye.

Laser speckle is generated whenever coherent to radiation is scattered from a surface whose roughness is comparable to or greater than the wavelength of the radiation. Wavelets from different depths of the rough surface interact with one another to give an interference pattern, not in the form of regular fringes as typified by Young's experiment, but as an irregular array of bright and dark speckles. These fill the whole of the 3D space where the wavelets combine, and it can be shown from statistical considerations that there will be a predominance of completely dark speckles, which explains the high contrast of the pattern and why it is so readily observable. The apparent size of the speckles is dependent on a number of factors which should become apparent in the course of the experiments to be conducted. For a more complete understanding of the phenomena see Dainty.

Subjective speckle pattern

Initial experiments introduce the student to speckle effects due to the observer himself, this is termed subjective speckle. The laser is mounted on an optical bench and its beam enlarged by directing it through a short focal length lens (e.g. a microscope objective) which focuses and then spreads it out to form a bright circular patch on a screen mounted at the far end of the bench. (Care should be taken not to look straight into the laser beam before it is widened out by the lens.) Keeping quite still the screen is observed from a distance, with one eye at a time only. The following experiments are carried out.

1. Screens made from materials of different surface texture, e.g. white card, coarse and fine sandpaper, matt aluminium etc, are introduced and the appearance of the speckle noted. Apart from differences in brightness no major change should be seen, showing that a speckle pattern is generated

independently of the roughness of the surface, provided that this is greater than the wavelength of the light.

2. With the observer stationary, the screen is moved from side to side. The motion apparently destroys the speckle. Moving the head from side to side, on the other hand, causes the pattern to shift. For students wearing spectacles, removing them changes the appearance but does not destroy the speckle pattern, i.e. it is not blurred out by defocusing. These latter effects offer a challenge to the student to explain, which perhaps can only be done satisfactorily on completion of all the experiments.

3. The polarization nature of the speckle is investigated by illuminating the screen with polarized light from the laser (introducing a polarizer into the beam if the laser is randomly polarized) and observing through a second polarizer. The difference between scattering from white card and from the aluminium surface is readily seen. This is explained by penetration of light into the card surface where multiple scattering takes place, so destroying the polarization. Single scattering by the aluminum surface does not do so.

4. The strong dependence of the speckle pattern upon the aperture of the imagining lens (the eye) is shown by viewing the illuminated screen through a series of pinholes made in black card with a needle. The smaller the pinhole the coarser will appear the pattern. When the pattern is viewed through a slit of variable width the speckles spread out into lateral streaks having a length inversely proportional to the slit width. These experiments show how diffraction at the

small lens aperture limits the detail to be seen to a size larger than the speckle. The speckle pattern formed by a camera lens equally has a graininess of its image that is dependent on the numerical aperture, or f-number.

5. Still viewing through the pinhole, the laser light is directed on to a hand held as stationary as possible. Disturbance of the speckle pattern is observed due to continual movement of the skin and subcutaneous blood vessels. Similar motion can be detected below the surface of ripe fruit, particularly red-skinned varieties such as tomatoes, where the laser light penetrates and is scattered internally by living cells.

Eyesight testing

The speckle rotation effect investigated above can be utilized in a device to test for the defects of vision of myopia (short-sightedness) and hypermetropia (long-sightedness). The test depends upon the physiological mechanism by which the human eye focuses upon a object. It does this by looking for an edge, changing focus until maximum contract is obtained when scanning across it. However, when observing a surface illuminated by laser light, the high-contrast speckle pattern existing throughout space is not localized in any one plane. The brain cannot then pick out a particular edge to focus upon and 'gives up', the eye reverting to its relaxed position where in a normally sighted person it would focus upon distant objects (focusing at 'infinity').

The eyesight tester, which is based upon a commercial instrument, consists of a slowly rotating drum illuminated by helium-neon laser light. This is observed 'at infinity' by the subject under test viewing it through a long focus lens mounted at its focal distance from the drum. As the drum rotates, an objective speckle pattern passes across the pupil of the eye. If the eye is focused at infinity, each speckle ray will fall upon the same region of the retina and the net effect will be a 'twinkling' for 'boiling' of the image. However, for a long-or short-sighted person the speckle rays will move up down the retina and as a result the whole speckle pattern will appear to move up or down. Distinguishing short from long sight is possible from the sense of the speckle motion, remembering that the brain inverts the upside-down image on the retina.

The eyesight tester constructed in this Department has a sand-blasted aluminium drum driven by a synchronous motor at a speed of one revolution per five minutes. Coaxial with this drum is a black and white striped drum which can be rotated at variable speed by a second motor. This is used to estimate the apparent speed of motion of the speckles as viewed by an observer.

Students are able to test their eyesight qualitatively by observing, with each eye separately, the apparent direction of speckle motion, whether up or down, and determining from the sense of drum rotation whether long or short sight is indicated. If the motion is to some degree diagonal it shows that astigmatism is present. Students wearing spectacles can observe the effect of removing them, while those with normal vision can simulate poor eyesight the aid of weak positive or negative lenses. If possible to obtain the use of a laser of difficulty wavelength, e.g. an argon laser, the chron aberration of the eye can also be demonstrated a person with near normal sight the illuminated with adjacent patches of green and light, will show different speeds of speckle motion.

Quantitative eyesight testing

The speed of rotation of the striped drum is varied to match that of the moving speckle. It be shown that the focal length the lens necessary to correct the observer's eye is given by

$$f_0 = -2\,\omega_0 Z^2 / \omega a$$

Where ω and ω_0 are the angular velocities of speckle and striped drums, a the radius of the stripped drum and *Z* the focal length of the viewing lens.

In practice it is not possible to match the with precision, but even so a measurement to half a dioptre for the correcting lens can obtained.

5

Keeping Moving to Stay Where You Are: Energy Flows and Steady States

Many systems need to be actively maintained to keep them in a steady state-centrally-heated rooms, living things, the Earth. The use of commercially available 'temperature sensitive film' allows qualitative ideas about steady-state systems to be easily investigated by pupils from lower secondary school onwards. Some examples of more advanced quantitative ideas which can be developed are given.

There is more than one way of staying the same. A bottle of red wine on a dinner table and a human being may each remain at an approximately constant temperature, but for very different reasons. The wine stays at the same temperature because it is at the same temperature as the surroundings. A human body needs to maintain itself at a temperature which is different from the surroundings; energy escapes due to the temperature difference between the warm body and the surroundings, and this must be balanced by an equal energy input from metabolism. Similarly, the earth moving around the sun and a car travelling down a motorway may both be moving at approximately constant speeds, but again for different reasons. The Earth's movements is essentially unchanging, while the car maintains a constant speed because the energy input (from the engine) balances the energy output (losses due to air resistance and friction).

So, some things stay the same because there is no tendency for them to change. Others stay the same because some difference (for example, a temperature differences) is actively maintained. A more direct example serves as a useful analogy, comparing a person standing still on the ground with a person walking up a 'down escalator'. The latter is literally a case of 'keeping moving to stay where you are'. There are many examples of such 'steady state systems'-a system which is maintained in the same state by a constant flow of energy or matter or both. These include simple domestic situations, biological systems, and large-scale phenomena, such as a centrally-heated room, a flame, a saucepan kept boiling, a living organism, an ecosystem, a tornado, the Earth, and so on Schrodinger noted that since living things are systems which constantly produce entropy, they need to be actively maintained (which they do by consuming 'negative entropy', for example, sunlight or food). Such systems need to get rid of entropy, and one way of doing this is by being warmer than their surroundings.

A steady-state demonstration

The apparatus show in figure 1(a) is a simple yet effective way of investigating a steady state system maintained at a temperature higher than its surroundings. It serves as a model of a centrally heated room, or of the human body of the earth. This 'steady-state box' consists of a small

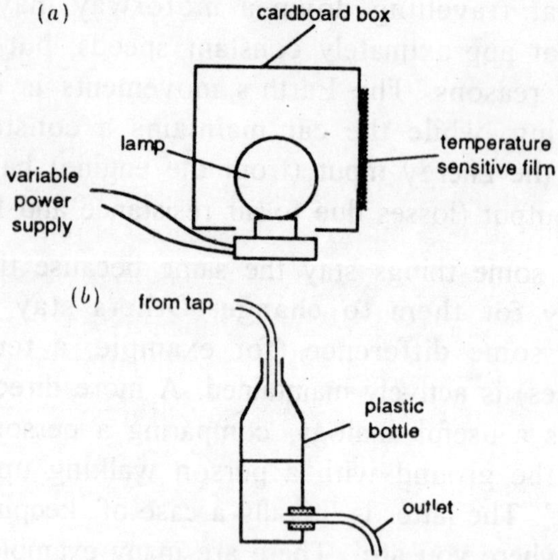

Figure: (a) A 'steady-state box'; (b) a 'Steady-state bottle'.

cardboard box with two holes. Into one hole is inserted a lamp (12V,24W would be suitable) connected to variable power pack; the other hole is covered with a piece of temperature sensitive film, which, as it warms changes colour from black to orange to yellow to green to blue. This material has recently become commercially available. It consists of a dispersion of cholesteric liquid crystals on a black background. The molecules in these crystals are oriented in a helix, the pitch of which is the same order of magnitude as the wavelength of visible light. The scattered light undergoes interference, giving rise to iridescent colours. Since the orientation of the molecules changes with temperature, the film changes colour as it warms.

Films for different temperature ranges are available; for the present purpose, the most suitable range is 25 °C to 30°C. (Temperature sensitive film can be obtained on small cards from the Science Museum Shop in London or in larger quantities from George Eliott, Stone Cottage, Tickerage Lane, Blackboys, East Sussex, TN22 5LX. At the time of writing the cost of a self-adhesive backed sheet 1' x 2' was E15). Adjusting the lamp to a low setting, the box can be made to remain at an intermediate temperature (i.e. the colour of the film stays green). This steady state is maintained by a flow of energy, the temperature of the box staying constant when the rate of energy flowing in balances the rate of energy flowing out of the box.

The major advantages in using temperature sensitive film for this demonstration is that, due to its low thermal capacity, the system responds rapidly, allowing changes in the energy inputs and outputs to be easily investigated. A breaker of warm water could also be maintained at a constant temperature in a similar manner with an energy input which balanced the rate at which energy escapes (only a few watts), but it would be very slow to respond to the small changes in energy input or output which would result in a new steady state being established.

A useful analogy for interpreting the behaviour of the 'steady-state box' is shown in the above figure. It consists of a plastic bottle with a tube at the top carrying water in from a tap and a tube at the bottom allowing water to drain into a sink. In this case, the steady state is maintained by a flow of matter, the amount of water in the bottle staying constant when the rate of water flowing in balances the rate of water flowing out of the bottle.

Changes to steady states

If the rate of input or output for a steady state system is changed, then the system will change until it reaches a new balance. For the 'steady state box', changing the input or output will change the temperature of the box; for the 'steady-state bottle', changing the input or output will change the amount of water in the bottle.

This idea is easily seen with the 'steady-state bottle'. Starting with an empty bottle, pupils could be asked to predict what will happen if the tap is turned on a little. The bottle starts to fill up, but it does not overflow, since as the water level rises, the output of water increases. Eventually the water reaches a steady level at which the input balances the output. Pupils could predict what will happen if the input is increased a little, or if the water flow at the bottom is reduced by putting a clip on the tube. The idea which needs some discussion with pupils is that we do not have to carefully adjust the input until it balances a fixed output-we can choose from quite a wide range of inputs, and the system will adjust itself until the height of water prducers exactly the output to balance the input.

Analogous changes can be explored with the 'steady state box',. This is best set up with the lamp at a lowish setting, and a hairdryer blowing cold air onto the film so that it is kept at intermediate temperature (green). Finding the right settings will need a little experimentation. The energy input can be changed by increasing or decreasing the voltage on the power pack-the film should change colour until it reaches a new steady state at a higher or lower temperature. Changing the voltage back to the original setting returns the box to its original temperature. The insulation may be changed by putting a cloth between the hairdryer and the box, the film becoming warmer; after removing the cloth the film returns to its original colour. The effect of changing the temperature difference between the box and its surroundings can be explored by turning the heater on the hairdryer on and off.

Some examples

Most manuals on central-heating installation have a table of *U* values which allows the degree of insulation of a room to be calculated. A *U* value is the rate at which energy passes (in watts) through a surface of area 1 square metre for each 1 °C temperature differences between one side and the other. Thus, the energy flow required to maintain a temperature differences between the room and the outside depends on the surface area of the room, its insulation and the size of the temperature difference. A typical U value for a fairly well-insulated wall is 2.5 W m^{-2} $°C^{-1}$. Taking for simplicity a rather atypical house consisting of a single room made only of walls, the surface area would be about 100 m^{-2}, and rate of energy loss would be about 0.25 kW for each 1 °C (2.5 W m^{-2} °C-1 X 100 m^{-2}). The power output from the radiator to maintain the room at a constant 20°C would depend on the outside temperature. On a warm day at 16 °C, the temperature difference would be 4°C, so a power output of 1 kW would balance the energy escaping and maintain the temperature difference; on a cold day at-4°C, the temperature difference would be 24 °C, and a power output of 6kW would be needed.

A car travelling down motorway is similar to a centrally-heated room in the sense that it warms the ground and air around it-directly by the hot gases produced by engine and indirectly by the car doing work against rolling friction and the drag of the air. The power loss due to rolling friction is not much affected by the car's speed, but the power loss due to air drag increases sharply with speed, as shown in table 1. It maintains a constant speed when the power output of the engine matches the power loss due to warming the air and the ground through friction. For example, with an engine output of 7kW, the car travels at a constant 70km h-1, at which speed the power loss is also 7kW. With a higher engine output, the car would accelerate until it reached a speed at which the power loss again balanced the power output of the engine. In fact, the input of energy to the engine does not always increase with speed, since the efficiency of the engine also increases with speed as the following table shows.

Speed (Kh^{-1})	Power loss: rolling fruction (kW)	Power loss: air drag (kW)	Total power loss (kW)	Effuiciency (%)	Fuel input to engine (kW)
50	3	2	5	8	63
70	3	4	7	15	47
120	3	12	15	20	75

The Earth, like a centrally-heated room is also maintained at a temperature warmer than its surroundings, with radiation from the sun on the earth balancing the energy radiated by the earth into cold space. The average radiation falling on the earth from the sun is 340 W m^{-2} though nearly a third of this (102 W m^{-2}) is reflected by the earth, with the remainder (238W m^{-2}) being absorbed. To maintain a balance, the earth's average temperature is such that an equal amount (238 W $^{m-2}$) is radiated. This can be estimated from Stefan's Law:

Power radiated per unit area = $\varepsilon\sigma T^4$.

Taking the emissivity, ε = 1 and the Stefan-Boltzmann constant, σ = 5.67 x $10^{-8}Wm^{-2}K^{-4}$, gives a temperature of

255 K or -18°C. This is rather colder than it actually is, the reason being that no account has been taken of insulation due to the greenhouse effect. In reality, energy is absorbed by clouds and by greenhouse gases in the atmosphere and radiated back down to the earth's surface. Thus, the radiation from the surface of the earth must be higher-in fact it is 353 W m-2. From Stefan's Law this gives a surface temperature of 281 K or 8°C-much more comfortable. Mars, with no atmosphere of greenhouse gases, is rather colder than earth at -60 °C, but Venus with a thick atmosphere containing water vapour and carbon dioxide is much hotter at 480 °C.

Of course there is more to the earth than just a warm rock in cold space. Its chemical composition, as well as its temperature, is kept away from equilibrium. Life processes maintain the presence of such molecules as oxygen and proteins on the earth. But like the temperature difference, life too is maintained by the flow of energy from the hot sun to the warm earth to cold space.

6

Isothermal and Adiabatic Measurements

The Adiabatic Gas Law Apparatus manufactured by PASCO scientific is a useful tool for measuring the pressure, temperature, and volume of a variety of gases undergoing compressions and expansions. Gases studied in this column represent the three general classifications: monatomic (argon), diatomic (nitrogen), and polyatomic (carbon dioxide). The apparatus is intended to qualitatively study adiabatic processes. An adiabatic process may be simulated by rapidly changing the volume of an enclosed gas, allowing little time for heat transfer between the system and its environment. Measurement of the pressure and volume of this gas can yield a value for the adiabatic gas constant, Y. If the volume is changed in small steps and if the gas is given sufficient time to come to thermal equilibrium after each change, measurements of pressure and volume will then correspond to an isothermal process. An ideal gas (one in which the pressure is not too large), regardless of classification, will have the pressure inversely proportional to the volume for an isothermal process.

Adiabatic processes

Thermodynamic theory gives a relation between the pressure and volume for a gas undergoing an adiabatic process as.

$$p_i V^{\gamma}_i = p_f V^{\gamma}_f = Constant \qquad (1)$$

where γ = ration of molar specific heat at constant pressure to the molar specific heat at constant volume = C_p / C_v. This ratio is equal to 5/3 for monatomic gases, 7/5 for diatomic gases, and 4/3 for "polyatomic" gases. In reality, γ varies for polyatomic gases depending on the complexity of the gas: for carbon dioxide the value is actually 1.30. In order to perform a linear regression on the data, we take the natural logarithm of both sides of Eq. (1). Rearranging terms, we obtain

$$\text{In}(p) = - \gamma \ \text{In}(v) + constant \qquad (2)$$

Thus a plot of In*(p)* vs In *(V)* for a given gas undergoing an adiabatic compression should have a slope equal to the negative of the ratio of specific heats. Since the thermal sensor lags behind the pressure and volume sensors, data on *p(T)* and *V(T)* were not used for measuring the adiabatic gas constant.

The adiabatic process is simulated by rapidly compressing gas in a cylinder for forcefully pushing down on a lever that is attached to a piston in the cylinder, whereas for an isothermal process the volume must be changed slowly.

Figure illustrates the adaptation of this apparatus to perform isothermal measurements. Gases are introduced into the cylinder by use of valves mounted at the bottom of the cylinder. It is important to keep the flow rate of gas

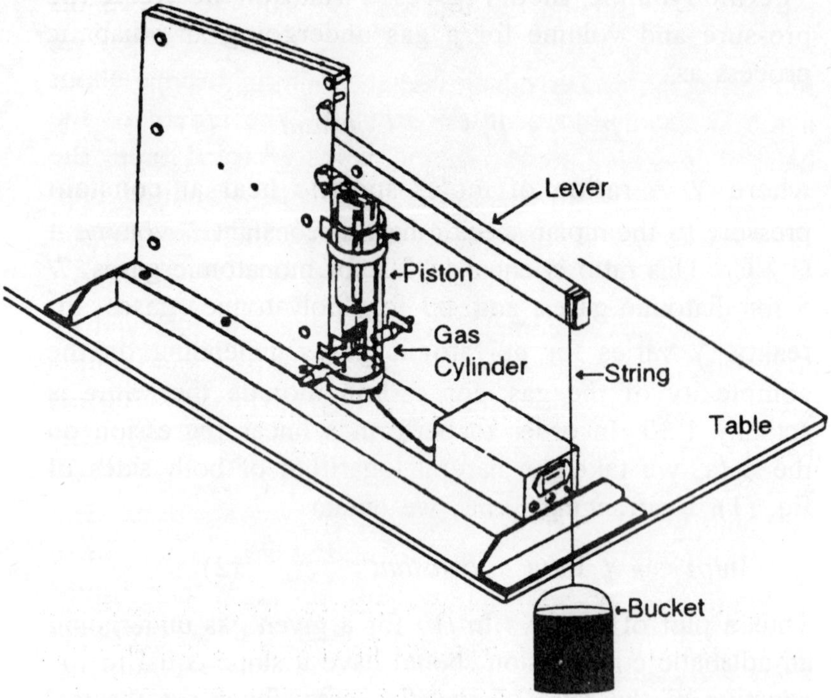

Figure: Schematic of Adiabatic Gas Law apparatus showing the addition of the bucket suspended from the lever arm for isothermal studies

very low; the temperature sensor uses very fine nickel wire wound on a frame near the bottom of the cylinder. If a gas is introduced into the cylinder too quickly, these wires could be damaged. To prevent this from occurring, we use modified Mylar balloons as gas storage devices. Many helium Mylar balloons found in stores have a "self-sealing" mechanism: a thin plastic sleeve inside the balloon that collapses after the balloon is inflated, keeping the gas inside. In order to access the stored gas, this inner sleeve must be removed. We found that inserting and subsequently removing a piece of rubber or Tygon tubing (approximately 3/16-in O.D.) will result in the inner sleeve being pulled outside the balloon. We cut the sleeve off near its end and removed it with fine tweezers. Then we reinserted the rubber tubing into the balloon, leaving about a foot of tubing outside the balloon. The mouth of the balloon is sealed to the tubing with electrical tape; the open end of the tubing is clamped with a bulldog clip. This device allows for long-term storage and easy removal of gas.

The adiabatic compression of the gas is achieved by forcefully pushing down on the lever. Pressure and volume data can be recorded during the approximately 100-ms period of the stroke by operating the data monitor software in "Oscilloscope Mode". The data are saved on disk and imported into a spreadsheet program. Raw data corresponding to the times after the stroke has been completed is deleted so as not to skew the linear regression. The spreadsheet then converts the raw data from the pressure and volume sensors into kPa and cc, respectively, using calibration data given by the manufacturer.

Isothermal processes

The ideal gas law states

$$pV = nRT \quad (3)$$

Taking the natural logarithm of both sides and rearranging terms, we obtain

$$\ln(p) = -\ln(V) + constant \quad (4)$$

This equation shows that the slope of a plot of In*(p)* vs In*(V)* for any classification of ideal gas is equal to -1. To create an isothermal process, we clamped the Adiabatic Gas Law apparatus to the edge of a table with the free end of the lever projecting over the edge. We hung a one-gallon bucket from the end of the lever, as shown in figure. The length of the string was such that the bucket is below the top of the table when the lever is at its highest point (the gas in the 'cylinder is at its maximum volume), and is short enough so that the bucket does not reach the floor when the lever is at its lowest point (minimum gas volume). With the software operating in "Monitor Input" mode, an option can be chosen to display the volume, temperature, and pressure data each second while only storing the data taken every 15s.

Lift the lever to its maximum height, and hold it for about 15s, to allow the gas to come to thermal equilibrium with the room. Initiate data-taking by the computer; after the first point is recorded, allow the lever to fall to a new equilibrium position with only the empty bucket attached. As may be seen from the raw temperature data displayed every second on the computer screen, the temperature of the gas rapidly decreases to near room temperature in about 10 s or less. After the second point is recorded, gently and quickly place a 0.500-kg mass in the bucket: the gas is compressed somewhat and shortly comes to thermal equilibrium at its newer and smaller volume and higher pressure. Repeat this procedure of adding a 0.500-kg mass each time, until a total of 22 data points are collected. (The process can be reversed: remove masses one at a time and then lift the lever for the final data point.) Since this process occurs slowly (in effect), any thermal energy produced by friction between the cylinder walls and the piston will have sufficient time to flow from the gas before the next data points is taken.

7

Low-Tech Solutions, High-Tech Results

Any number of physics instructors have reported on curricula that are designed to encourage active rather than passive learning. These techniques emphasize the student's role in creating relevant personal meaning that is centered on concepts, as opposed to memorizing detached facts or frequently used algorithms. Many of these curricula utilize Microcomputer-Based Laboratory (MBL) tools or other technology-dependent features. Others rely on decidedly "low-tech" techniques. But they all share a strong dependence on interaction through language.

As a teacher in a large suburban high school, my situation required a low-tech approach that consumed less money than MBL tools yet produced significant conceptual gains. I report here on the preliminary results of a language-rich teaching method that helps students learn physics concepts. This method features the Active Mental Processing (AMP) journal.

Active Mental Processing(AMP) Journal

Introduction

Many people learn more effectively when they process information actively rather than passively. Examples of active processing are:

- thinking ahead
- anticipatingactions
- interacting
- pondering implications
- predicting results
- discussing
- seeking applications
- self starting
- posing and answering question
- recording detailed notes

Examples of passive processing are:

- thinking that attendance is equal to learning
- waiting for the teacher to initiate action or dialogue
- circumventing thinking in favor of rote memory or algorithmic problem-solving strategies.

Passive students do not perform well in classes that require higher level mental processing since they are not accustomed to engaging the material activity. The AMP journal will train such students to actively process knowledge.

Rules

1. Notes of the day (2points per school day possible)

Purchase and maintain a spiral notebook to hold a DAILY journal entry plus two other requirements. The bulk of the DAILY entries will be class notes. This is not just what I wrote on the board that day, but includes what I did, what we talked about, what you did, as well as important things that I said or emphasized (format below). I will frequently give you time in class to write "notes to yourself" in order to consolidate and solidify a key point of the day.

Please include these comments. If you are absent, provide notes from your reading of the text.

2 points-This goes beyond copying the notes on the board. Please include additions and embellishments that help you to understand and remember the meaning of the activity or discussion. For example, you should include captions to any graphs, explaining the significant of the data and a reminder of how the data was collected. Also, when we have open lab days or work days, please tell specifically what you did. For example, if you worked on chapter problems, include that work in the entry. If you worked on a project, tell me about it. Please give good details such as providing a sketch or summary of specific actions or decisions. Solve all chapter problems in this "Notes" section.

1 point-This results from simply copying the notes with little evidence of personal interaction with the ideas or problems of the day. Reproducing a sketch without any caption or explanation would be an example.

2. Question of the day (2 points per school day possible)

I expect you to ask yourself at least one question about physics DAILY. This will usually refer to the topic of the day, but can also come from the text, newspaper, or other activities. You must ask a different question each day Purchase a RED pen and write this question in red toward the bottom of your journal entry DAILY. Give your answers based on your knowledge at the time. Keep the questions simple and sincere. Trivial questions such as "How do I spell Physics?" will not receive credit. Do not wait until the last moment to concoct several days' worth of questions. I can generally discern this and will not award credit. Here are various examples.

2 points-"From where did that vector pointing to the right on page 42 come? I think it might be the horizontal force, but I am not sure".

2 points-"You said that the speed of light is the 'speed limit' of the universe. If two objects are traveling toward each other at 90% the speed of light, then what is the speed of light from the opposite object? I think it has to be greater than the speed of light".

1 point-This question will sound trivial or contrived. It will not have an answer that indicates true involvement. For example, "How many positives add to how many negative charges to make neutral? Same," Students at your level of physics are expected to know this.

3. Application of the day (2 points per school day possible)

Think about, invent, or notice an application of the physics concepts mentioned in your journal entry and write a short explanation of how it is an illustration of the day's topic. If a specific examples is difficult, create a metaphor, analogy, or story that illustrates the main concept of the day. Purchase a GREEN pen and write this in your spiral notebook DAILY (format follows).

Please be realistic. Examples follow.

2 points-Personal history and stories are the key to this score. Please think of situations in which you actually participated or at least could easily imagine participating. For example, "One day my Mom was baking my little sister's birthday cake and I had to vacuum the carpet. The blender was going and all the lights were on. My cat's toy mouse got clogged in the vacuum cleaner and the fuse blew. This is an example of too much current being drawn by the circuit at one time."

2 points-"In class we talked about inertia. When I was 'rear-ended' last month, inertia caused the screwdriver in the glove box to fly out and land in the front seat. It was at rest and 'wanted' to stay at rest as my car was bumped forward."

1 point-These applications lend to sound global without much connection to specific every day situations. For example, "Energy is conserved in everyday life."

Please feel free to ask me to read and evaluate any question or application before the due date. I will do my best to guide your active mental processing.

Cautions

1. The AMP method is most valuable when students are current with their journal entries at all times. I reserves the right to spot check these journals without notices. Students will lose the points for all entries not up-to-date.
2. It is the complete responsibility of the student to obtain class notes when he or she is absent. A DAILY entry must be recorded regardless of absence of the day's activity. A student may substitute notes from text reading for class discussion notes. The general topics should match. Students with excused absences have one extra day per absence to show me their journal.
3. A student may not copy another student's question of the day or application. These portions of the journal must be unique. No points will be awarded if this is not the case. The only valid reason for copying class notes is absence. other-wise, journal entries must reflect your personal learning experiences and mental processing.

4. Losing your red or green pen for a day does not excuse you from your journal responsibilities. If this happens, separate the question of the day and the application from the notes, give an appropriate heading to each section, and rapidly replace your pens. (The color coding helps me to grade these assignments efficiently.)

I give student a detailed description of requirements for the AMP journal on the first day of class. The format is very structured and requires students to make three different entries for each school day. These entries are designed to foster students' conceptual organization and facilitate grading.

Students' notes should contain more than what I write on the board. Over the years, I have found that few students take notes and those that do write only what is written on the board, with little personal interpretation or elaboration within the context of the discussion. I train students how to listen for key phrases and important ideas made during our classroom discussions and how to personalize them. I accomplish this by stopping my presentation at crucial junctures and asking students to write what they think. Frequently I end these one-to-three-minute sessions with a verbal recap against which students check their notes for accuracy and completeness.

More important, I structure students' conversations about physics concepts by giving them something interesting to talk about. Classic physics demonstrations framed in relevant settings form the basis for many of these conversations. The cycle is usually prediction, observation, then making personal meaning. Important ingredients of the AMP journal technique are "turn and talk" as well as "write yourself a note" times. The key is time. If I do not give students time to talk to each other about their beliefs, predictions, and interpretations, they attempt to memorize my ideas rather than creating their own. During students' conversations, I monitor discussions and judiciously guide students who are at an impasse or who demonstrate conceptual misunderstanding. Furthermore, I make mental notes of the most common fallacious phrases, which I make a point to contrast against accepted use during the lesson summary. As a result, students slowly replace their naive use of physics words by emulating the correct use, which I model in recap sessions. Students translate classroom events in their notes by individual interpretation, not by transcription of the teacher's words.

As shown above students record one question and one application per day. During the first few weeks of class especially, I must help students with these. So I make sure that during each lesson I give several examples of appropriate questions and applications. Often I start the day by writing a question on the board: this serves as the theme for the day. I pepper my talk with applications of the topic at hand and draw attention to the fact that I have mentioned an application. Otherwise, students do not realize what they have just heard. At the end of the period, I leave 2 to 4 minutes for students to write a question and application in their journals.

Practical matters

Encouraging students to verbalize their conceptual understanding of physics in a nonthreatening classroom environment is the core of the AMP journal. Managing students' conversations in order to foster their active construction of meaning is theoretically sound, but if this effort were to take too much teacher time, then the method would not be practical. I suggest the following "rules" to maximize the student benefit and minimize the teacher time associated with the AMP journal.

Format: Establish a strict format with key ingredients in an easy-to-follow structure. I mandate that the question and application of the day be in different colored pens and be recorded after the day's notes. Also, I ask students to write on every other line to ease eye strain and allow room for teacher comments.

Grading: Publish and discuss with students the grading criteria and give examples the first day of class. Pick up and grade the journals every three weeks. With this pattern, I spend approximately one hour per class of 25 students per grading session. Do not grade for correct answers and to not write answers in the journals. Instead, make a point of addressing journal questions in class. The chances are great that other students have the same questions. Make comments of general encouragement in the application section and mention a few of the more memorable ones in class.

Time: Give students time to think, write, and talk about physics when using the AMP journal technique. This method does not work when students feel rushed, in a testlike environment, or if they believe their comments will be ridiculed or discounted. Patiently model proper use of physics words and respectful exchanging of ideas to foster naturalistic conceptual discourse.

Assumptions: Do not assume that high-school students know how to take meaningful notes or translate school physics into their everyday world. Telling students concepts does not work.

Results and observations

Scores on the FCI before and after I introduced the AMP journal are significantly different. In school year 1993-94, the posttest mean was 14.9(51.4%) correct and in 1994-95, with the implementation of the AMP journal, the posttest mean was 19.1(65.9%). In 1994-95, the pretest mean was 8.38 (28.9%) translating to a gain score of 36.9% points. I have no pretest measure for 1993-94. Therefore initial group differences could explain the result, although my classes have been quite homogeneous for 10years on measures such as mathematics background and previous science success. Hestenes et al, suggest that neither mathematical nor socioeconomic background affect FCI posttest scores, and since pretest scores on the FCI for beginning students are uniformly low, posttest scores are the only measure needed to evaluate students or teachers. Furthermore, Hestenes suggests that a gain score of 36% points is impressive and that 60% correct responses represents a conceptual threshold indicating emerging Newtonian thinking. Pretest versus posttest scores on the six conceptual dimensions that Hestenes suggests are tested by the FCI.

Huffman and Heller determined through factor analysis that the FCI actually tested only two factors. Factor one focuses on Newton's Third Law and includes test items 2, 13, 11, and 14 while factor two involves kinds of forces and includes test questions 5, 9, and 22. Figures 4 and 5 show how my 1994-95 students performed on these factors, compared with the 1993-94 classes (which did not keep an AMP journal).

My students continue to perform poorly on several items of the FCI those questions for which my students scored less than 60% correct. Though they performed well on Newton's Third Law and kinds-of-force questions, they showed room for growth in the independent nature of horizontal and vertical motion, "overcoming" forces, kinematics, and impetus.

As I reviewed my lesson plans for 1994-95, I noticed three differences between the presentations made during the factor 1 and 2 concepts and those concepts represented. One, I strictly followed the AMP journal methods outlined earlier. I made a point of letting students have time to think and struggle with a number of counterintuitive situations in the AMP format of discussion. This included students' personal resolution of conceptual conflict accompanied by time to record thoughts and consolidate ideas.

I have observed that students listen and perform differently under the AMP teaching method. They listen for and write down applications as they hear them in class because they know they must have at least one example of each application per day. I see questions and applications scribbled in journal margins indicating keen attention. As the year progressed, I saw gradual improvement in students' ability to express accurately physics concepts, both in writing and verbally. Their notes became elaborated with personal comments, indicating individualized ways of understanding the day's topic or concept. Finally, I saw students' memory being affected. One student commented that the only reason she remembered all the concepts was because she had to come up with applications every day. For her, connecting school physics to her personal world helped her internalize knowledge.

8

The Automotive Ignition System

An automobile engine runs on gasoline, but nothing happens until the gasoline/air mixture is ignited. The ignition system is designed to provide that service by causing a spark to jump across a gap in the spark plugs at the appropriate time in the engine's cycle. This note discusses the Kettering ignition system used on all vehicles for about a half century, and then some transistorized ignition systems. For purposes of simplicity, many circuits are simplified and numbers are approximate.

The Distributor

In a multicylinder engine, the pistons do not move up and sown at the same time, and therefore the expansion strokes, when the fuel burns and expands, are not simultaneous. One purpose of the distributor, the reason for its name, is to distribute the spark from a high-voltage transformer, known as the coil, to the spark plugs at the appropriate time. Figure 1 shows a high-voltage wire from the coil to the center of the distributor cap.

From this high-voltage wire, a rotating connection (brass-to-graphite, usually) is made to the rotor, which is

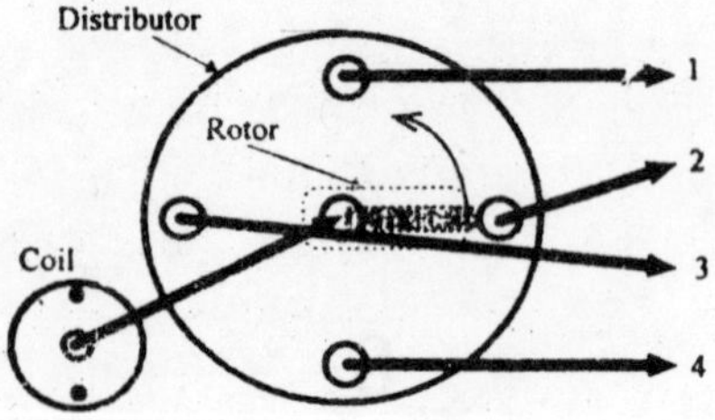

Figure 1: Top view of the distributor. High voltage is selivered from the coil to the ceter terminal of the distributor

rotating at half the rate of the engine proper. The rotor has a metal piece that makes close contact (usually brass-to-brass) with stubs in the distributor cap, from which high-voltage wires lead out to the various spark plugs. In fig.1, spark plug 1 fires first, then 3, then 4, then 2, this being the *firing order* for a four-cylinder engine with all cylinders in line. The reason the distributor shaft turns half as fast as the crankshaft is the piston must go up and down twice to complete the four-stroke sequence.

Inside the distributor reside the points (switch contacts) and the condense (a capacitor). As the *distributor shaft turns, the distributor cam* opens the points a distance of about 0.4 mm (0.016 in) by pushing on the *rubbing block*, an insulating material. The switch contacts of the point set are made of tungsten. In reality, a spring (not shown) pushes the moveable point into contact with the fixed one, and the points do not open until the rotating cam forces them open; at that instant, the spark occurs.

A wire from the coil is attached to the moveable point, and also to one terminal of the condenser, the case of which is metal that is grounded to the chassis.

The distributor also contains (usually) two mechanisms for *advancing the spark*, that is, causing the spark to occur earlier (in angle, not in time) at high engine RPM. To describe them in detail would deflect us from the electronic details that we wish to emphasize in this note.

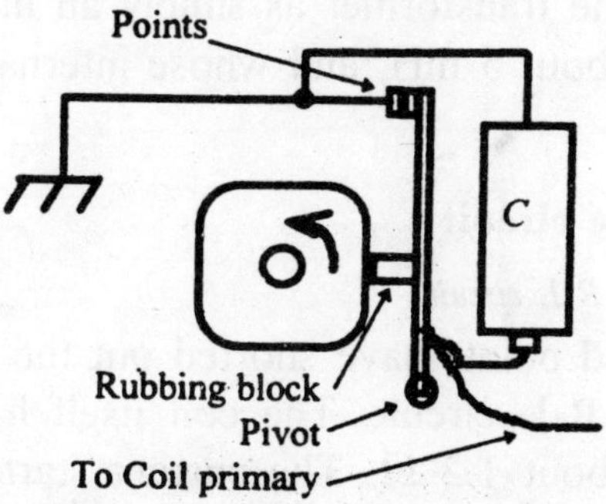

Figure 2: Points and condenser. One terminal of the points and one capacitor plate are tied to the chassis of the car

The Kettering Ignition System: Basic Circuit

Figures 1 and 2 are typical of those found in automotive repair manuals, in that they show parts layout. Of more interest to a physicist is the circuit diagram and explanation of its physical principles.

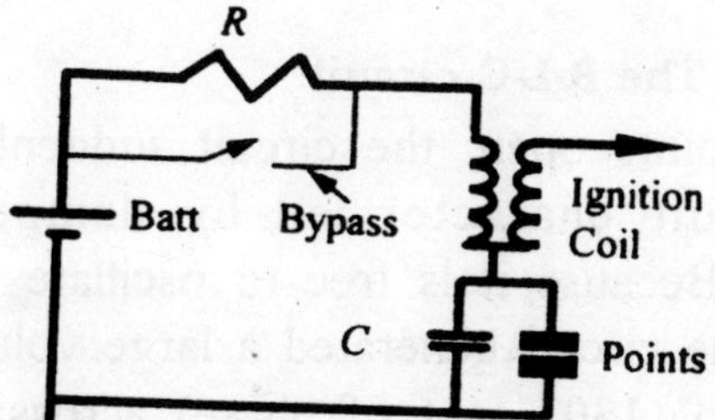

Figure 3: Ignition circuit

We show such a diagram in fig.3, which depicts the battery, a resistor, and a transformer in series when the points are closed. The resistor is either a discrete unit mounted on the firewall, or a resistive wire in a wiring harness. When the points are open, the capacitor is in series with the other circuit elements. For the time being, let us regard the transformer as simply an inductor whose inductance is about 5 mH, and whose internal resistance is about 1.2.

Behavior of the circuit

Points closed: The R-L circuit

Since the closed points have shorted out the capacitor, we have a simple R-L circuit. The coil itself has an internal resistance of about 1.2 Ω. The current starts at zero and builds up toward its saturation value of $I = V_{batt} / R_{top}$, where the time is measured from the instant the points close. The current is shown as a percentage of its final value.

The energy stored in the coil increases until the saturation value is achieved, or until the points open. In fig.4. we have used $R_{tot} = 2.2\ \Omega$ and $L = 4.6$ mH, which are actual values for parts in my cache of automotive necessities.

Points open: The R-L-C circuit

When the points open, the circuit suddenly becomes an R-L-C circuit characterized by damped sinusoidal oscillations. Because it is free to oscillate at its resonant frequency, the circuit generated a large voltage across the coil, and also (180° out of phase) across the capacitor, typically 300V.

However, things conspire to keep that high voltage

from happening. The coil is a transformer of about 85:1 turns ratio (which for numerical simplicity, we will round off to 100:1). If we have 300 V in the primary, then we will have 100 x 300 V = 30,000 V in the secondary, this being the voltage fed by the distributor to the spark plugs. When the spark occurs, the circuit is no longer a simple R-L-C circuit.

Effect of the spark

When a voltage is impressed across the circuit, no current exists until some critical voltage is reached, after which the voltage rises no more, but the current increases. This is also why fluorescent tubes require a ballast in the circuit-to limit the current. This behavior is common to neon advertising signs, stroboscopes, discharge lamps, and many other devices in which the current path is through a gas. The rapidly rising portion of the curve was used as a constant-voltage reference in many old electronic circuits before Zener diodes became available.

The spark between the electrodes in a spark plug obeys similar behavior, in suddenly occurring at some given voltage, and then keeping the arc voltage nearly constant.

The sparking voltage of the spark plugs depends upon the gap between the electrodes, the temperature of the air/fuel mixture in the region, the temperature of the negative electrode (which actually undergoes thermionic emission), and the cylinder pressure at the instant. Importantly, it is far less than the 30,000 V that can come from the ignition coil.

Other considerations

Energy particulars

When the spark plug fires, it draws energy from the

ignition circuit. Suppose if fires at 8000 V. This limits the voltage in the primary to 80 V (instead of its "expected" 300 V), and drains away the energy. Therefore, the apparent R-L-C circuit does not undergo the typical damped oscillations, but rather goes to one peak and then settles down immediately, with no oscillations and with the battery voltage appearing across the points.

Two important things happen at high engine RPM. First the voltage required to cause a spark in the spark plug rises. Typically, 7000 V is adequate at low RPM, but 10,000 or more may be required at high RPM. The *available* voltage, however, decreases. At high RPM, the points open and close more frequently. Especially, the length of time the points remain closed between sparks becomes smaller, the current is always much below the saturation current, and the energy stored in the magnetic field is correspondingly smaller than its own saturation value. For example, on an 8-cylinder racing engine running at 8000 RPM, there are 8000 (8/2)/60 = 533 point openings per second, or 1.88 milliseconds between sparks. The points are closed only two-thirds of that time, so they remain closed only for about 1.25 ms. Only about 12 kV is developed, and only about 20% of that energy is available at low RPM. For this reason race cars often use a dual ignition system.

Purpose of the condenser

It is often said that the purpose of the condenser is to keep the points from "burning" (arcing, really). To see why this is so, imagine the behavior of the R-L-C circuit with $C \approx 0$. As the points open, the current changes from (say) 5 A to zero very abruptly: the induced voltage across the coil (L di/dt) is very high, and so is the voltage across the points. If the points haven't moved very far apart, there will be an arc across the points to the detriment of the points. Worse yet, the energy initially stored in the coil is dissipated in the point gap instead of he spark-plug gap.

Starting conditions

When an engine is running, the alternator output voltage (hence the voltage available to devices in the car) is about 13.8 V, above the 12.6-V open-circuit voltage of the battery. The ignition system is designed to run at the 13.8 V.

When the engine is being started, the battery voltage drops well below 12.6 V because of the large current required to run the starter. At this time, the spark-plug electrodes are cold and require a higher voltage to spark then when the engine is warmed up. In order to provide adequate ignition voltage when the car is being started, the 1- Ω resistor is by passed, putting the battery voltage directly onto the terminal of the coil. Indeed, the resistor is put into the circuit so that it can be bypassed. (If the by pass switch fails, the engine runs well when started, but is hard to start. It seems to "catch" only when the starter switch is just released).

Modifications to ignition system

The Kettering ignition system dependably fired spark plugs in automobiles for about five decades before any significant improvements were made. The main problems with the system was the points. Eventually, the points, which were initially shiny pieces of tungsten, would become pitted. Also, at times, lack of lubrication for the rubbing block could cause the plastic device to melt and then to adhere to the cam, causing rapid wear of the remaining block. Wear in the distributor shaft would result in wobble, causing the points to open at unpredictable times.

It was apparent from the beginning that transistor circuits would be useful, but transistors capable of handling both the high voltages and the high temperatures of the engine compartment were very expensive, if and when they were available. The early transistorized and capacitive-discharge units were usually packaged in large hear sinks. These were after-market devices intended to be used with the points serving as a triggering mechanism.

The capacitive-discharge ignition system. One clever modification of the ignition system is to store the energy electrostatically in a capacitor $E = 1/2CV^2$ rather than magnetically $E = 1/2LI^2$ in the coil. In the capacitive-discharge ignition (CDI) system, a transistorized oscillator is connected to a transformer that boosts the voltage to about 300 V. This voltage is rectified and used to charge a capacitor. The capacitor voltage is isolated from the ignition coil by a silicon control rectifier (SCR).

The SCR acts like an open switch until a signal has been sent to its gate, after which it acts like a rectifier. It remains in the conduction mode until the voltage reverses, at which time it turns off.

When the points are closed, the gate of the SCR is grounded. When the points open, the voltage on the gate suddenly increases, and the SCR turns on, dumping the energy stored in the capacitor into the ignition circuit. The sudden surge of current through the coil causes a high primary voltage to be induced, and the high-voltage output in the secondary is produced.

The primitive transistorized system. Early transistorized ignition systems continued to use the points as a trigger. The resistor placed in series between the primary of the coil and the points limits the current through the points. When the points are closed, transistor Q_1 is turned off and transistor Q_2 is turned on ; Q_2 serves as a substitute for the points, shunting the condenser. When the points are open Q_1 is on and Q_2 is off.

The modern transistorized system. Both the CDI and the primitive transistorized systems have the advantage that the ignition points carry only very small currents, typically less than 0.1 A, but they have the disadvantage that the ignition points are still part of the circuit: everything depends upon a mechanical switch that has to open and close several thousand times for every mile of travel. The weak part of the circuit remains.

Given that we wish only to apply a periodic electrical impulse to initiate a spark, there are several nonmechanical ways to accomplish the task. For example, we might use a light source a slotted disk. Such devices are used as laser choppers in laboratories and as signaling devices in some computer mice, but they can be foiled by dirt. That is not a good prescription for an automobile.

Therefore, most automobile manufacturers use some sort of magnetic pickup system. One way to do this is to have a rotating system of iron wedges that pass by another iron wedge that is magnetized by a current-carrying coil, as shown schematically in the following figure for the system used in general motors automobiles. Another way is to have an electromagnet (Chrysler). The magnetic field suddenly increases when the points pass by.

The sudden change of magnetic field may be

measured by a Hall probe, by the induced emf in the coil providing the magnetizing current, or by the induced emf in a second coil. The many schemes used would require too much space and considerably more knowledge than I possess. Invariably the circuitry is encased in epoxy, and the diagrams in repair manuals show only parts layout (hook wire "A" here) rather than real circuit diagrams.

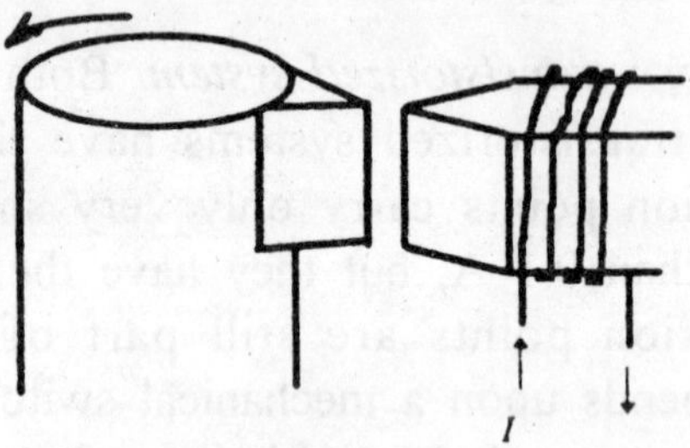

Figure 4: Rotating-wedge system

Classroom observations

In many experiences, students enjoy demonstrations of the conventional ignition system, if only because of the sparks. All electrical connections are accessible for measurements with meters and/or an oscilloscope. The reader whose cache is not as well stocked can readily obtain parts—durable and inexpensive ones—from any "Used Parts Emporium". It is easier to get a ceramic ignition resistor than to find (and explain) the resistive wire among many normal ones in a wiring harness.

9

The Fabrication of Integrated Optical Glass Waveguide Sensors

Optoelectronics is the synthesis of optics with techniques from the electronics industry and is now rapidly expanding within the areas of sensors, telecommunications and signal processing. Integrated planar optical waveguide sensors are extremely sensitive to variations in sample environment such as: refractive index, absorption, or fluorescence. Changes in these optical parameters may cause rapid modulation of light travelling within shallow glass optical waveguides, offering the possibility of continuous on-line integrated sensors for industrial and clinical applications.

The ability of glass to transmit or 'guide' light has been known for many years. However, it is only recently that it has been possible to make specialist glasses of exceptional purity and transparency and to use them design optical waveguides and fibres that are capable of surviving realistic working conditions.

All guided optical sensing techniques are based on the total internal reflection of light. When a light ray inside a transparent material approaches the interface between tat material and another transparent surrounding medium, the ray is internally reflected if it approaches the surface at an angle greater than the critical angel, Θ_c. For any interface $\sin \Theta_c = n2/n1$ where $n_1$is the refractive index of the waveguide material, n_2 is that of the surrounding or cladding medium and $n_1 > n_2$. Now, the electromagnetic field of the reflected light actually extends (or penetrates) of the order of a wavelength into the surrounding fluid medium, and may interact with the fluid in this region. Hence, if an optically absorbing material is placed in contact with the fibre cladding, the electromagnetic wave travelling in the waveguide will be attenuated. Several traditional optical fibre devices used for sensing function on this basis.

The sensitivity of total internal reflection spectroscopy of a fibre may be improved by increasing the field penetration at the overlayer. To achieve this, the fibre core is exposed by polishing. To enhance the field interaction further we may increase the length of core exposed. The polishing process produces a flat surface on an otherwise curved fibre which is useful for chemical and thin-film deposition. However, the polishing process is labour intensive and time consuming which is a serious limitation of the mass production of devices. Thus planar waveguides offer the advantages of inherent surface flatness, long path length and direct access to the exposed waveguide 'core' in the surface of the glass.

In a planar waveguide, as in a fibre system, the peak of the electric field is close to the centre of the waveguide with the 'tail' of the field penetrating into the adjacent medium. Planar sensors differ only in the way we harness the electric field's penetration of the surrounding medium. To make specific sensors, it is necessary to deposit chemically active films, metal and dielectric layers, or apply electric fields between the metal layer and an electrode in solution. Any changes in the optical properties of the thin film or surrounding superstrate will affect the attenuation of the incident input optical wave.

Fabrication is without doubt the most important process in the manufacture of optical waveguides. Present planar waveguide devices for sensing applications are formed in either cheap soda-lime glass or BK-7 glass substrates by a simple fabrication process. The fabrication process may be readily tailored to meet differing design specification.

Typically the process involves:

(i) *Photolithography*-the etching of channels in an aluminum mask on the glass substrate.

(ii) *lon exchange*-glass samples are immersed in a molten salt bath, usually maintained at several hundred degrees Centigrade, for say, an hour lons in the exposed glass are replaced by other irons in the molten salt, creating a region of either higher or lower refractive index in the exposed surface compared with inside the bulk glass. By placing a second photolithographic mask on top of the waveguide and immersing the sample in a different molten salt, it is even possible to add further features, such as a periodic grating in the waveguide, creating a zebra-crossing-type effect!

(iii) *Polishing*-many waveguides on several substrates may be end-polished together at once (this is necessary to achieve efficient light coupling from an optical fibre, or lens, into the sample) and the metal mask then removed.

(iv) *Evaporation*-a thin film, such as metallic silver, may be evaporated either directly onto the waveguide or onto a second 'sandwiched' layer, already deposited on top of the waveguide.

One recent exciting development is the fabrication of an integrated optical waveguide adsorption sensor. Light is coupled into a waveguide of about 4 microns width into which a thin silver film has been deposited. By the application of an electric field between the film and fluid, it is possible to control the adsorption and desorption of charged chemical species onto the waveguide surface and hence change the attention of the transmitted beam through the waveguide. The attenuation is enhanced by the transfer of energy between the waveguide mode and a lossy surface wave existing at the metal/fluid surface. If the chemical composition at this surface changes then the coupling of energy between the waves will alter significantly. As the negative voltage applied to the silver film is increased further, more positive ions are attracted near to the film; this results in even greater changes in transmission, observed within only a few seconds. This kind of device will allow rapid in situ determination is intended to be the monitoring of pesticide levels in drinking water.

IBM is currently looking at increasing the number of different sensing elements on a single substrate by decreasing individual element size. The printing of patterns on surfaces using light is a key element of optical processing, and the principles have changed little since their inception with semiconductor devices of the sixties and seventies. Current ultraviolet mask exposure routinely achieves a sharp image resolution of about 1 micron. Image sharpness on a substrate is controlled by light diffraction as it propagates through the mask aperture. The greater the amount of diffraction, the more 'blurred' the image on the substrate will be. The amount of diffraction is related to the wavelength of light, λ, used (diffraction ∞ λ), and hence a reduction in the exposure wavelength will improve image resolution. To achieve sub-micron resolution, IBM in New York and Vermont is examining the use of short x-ray wavelengths between 0.5-4.5 nm.

10

Basic Electricity– A Novel Analogy

Many introductory physics tests use the analogy of water flow when introducing concepts in basic electricity. While such visualizations may be of assistance to the students, the analogy would be difficult and messy to realize as a practical demonstration, as well as being fraught with other complications such as hydrostatic pressure confusing any link between pressure and potential.

I feel that preoccupation with a water analogy may have distracted teachers from considering air as the fluid analogous to charge in a demonstration with air flow analogous to current. I have set up an easily realizable demonstration that enabled first-year students being introduced to the concept of electric current to grasp the relationship between current, voltage, and resistance immediately. With the apparatus in front of them, the students were also easily able to make predictions as to how changing one variable, while keeping the other constant, would affect the third. The apparatus is shown in fig.1. It was assembled from readily available components in a few minutes.

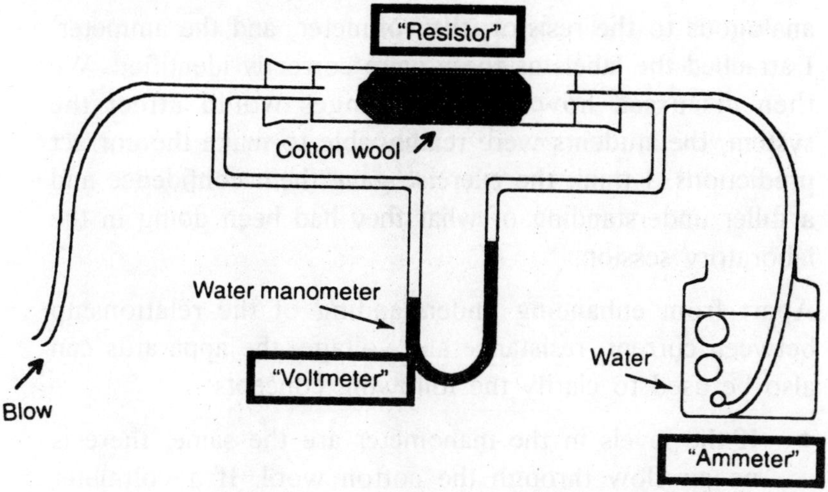

Figure 1: Apparatus for demonstrating the relationship between current, voltage and resistance

Introduced the apparatus to students in a laboratory session, after they had performed an exercise designed to familiarize them with an ammeter, voltmeter, and variable power supply. They had plotted graphs of current vs voltage for a flashlight bulb and a resistor. I think that at this stage many students found the circuit to be a confused mass of wires on the laboratory bench, and the current and voltage were two numbers that they got from the instruments.

These numbers did the "right thing" when a certain graph was plotted, but the students lacked an intuitive feel for what was going on in the circuit.

I presented the apparatus as depicted in fig.1 without labels, but told the class that it was a simple analogy to what they had been doing. I blew into the rubber tube and drew attention to the difference in levels in the manometer and the bubbles in the jar. I then asked which parts were analogous to the resistor, the voltmeter, and the ammeter. I attached the labels as these were correctly identified. We then discussed how various changes would affect the system; the students were readily able to make the correct predictions. I think the exercise gave them confidence and a fuller understanding of what they had been doing in the laboratory session.

Apart from enhancing understanding of the relationship between current, resistance and voltage, the apparatus can also be used to clarify the following concepts:

- If the levels in the manometer are the same, there is no air flow through the cotton wool. If a voltmeter across a resistor reads zero, there is no current in the resistor.
- There is no air flow through the manometer. An ideal voltmeter draws no current and is only measuring the potential difference across the resistor.
- Air flows from the high-pressure side of the cotton wool to the low-pressure side. Electric charge in a resistor flows from high potential to low potential.
- Air leaves the month of whoever is blowing and enters the apparatus at the highest pressure. Electric charge flows from the high potential end of the energy source towards the low potential end.
- Bubbles form in the jar almost as soon as someone starts blowing. Electric charge starts flowing in all parts of a circuit almost simultaneously.
- The air has to flow through the jar and the current manifests itself in bubbles per second. This air cannot accumulate, but is part of a complete circuit (eventually leading back to the mouth of the blower).

An ammeter has to be placed in the circuit, and may itself impede the current.

When circuits get more complicated, there will be exceptions to these arguments, and the type of analogy presented here will break down. Students will, in time, be expected to cope with such situations, but they will only be able to do use once they have a firm grasp of the basic concepts of electric current and are able to use them.

11

Demonstrating Reduced Gravity

The effects of gravity are so inherent in life on Earth that we often ignore how these effects complicate and sometimes mask fundamental processes in chemical and physical systems. To highlight the importance of gravity, we built a series of experiments, approximate for the classroom, to demonstrate how the behavior of common physical systems change when gravity is nearly eliminated.

Based on Newton's law, $F = ma$, Einstein realized that a body of weight $W = mg$ while stationary on the Earth, would appear weightless if it were in free fall. Imagine that the body is within an imaginary box (i.e., frame of reference) that is failing at the Earth's gravitational acceleration ($a = g$). Relative to the falling frame of reference, the body is motionless, there is no net acceleration, and therefore $W = 0$. The state of free fall is often referred to as weight-lessness or microgravity.

In several research labs around the world, scientists take advantage of this idea and perform "gravity-free" experiments in drop towers, airplanes, and spacecraft.

Drop towers typically provide several seconds of microgravity and, as the name suggests, operate by hoisting the experiments to the top of a tower and dropping them. For example, NASA Lewis has two towers: one provides a fall distance of about 25 m and the other provides 130 m. These fall distances correspond to 2.2 and 5.2s of microgravity time, respectively. The Japanese Space Environment Utilization Center in Hokkaido, Japan has a 10-s drop tower with a fall distance of about 500 m, roughly one-third of a mile. Similarly, free-fall experiments are performed aboard airplanes, sounding rockets, and the Space Shuttle by flying a trajectory such that the craft falls freely towards the Earth (and around it in the case of the Space Shuttle).

Using this same principle, we developed a miniature drop tower at NASA Lewis, named the Reduced-Gravity Demonstrator, to illustrate the effects of gravity on a variety of phenomena including the way fluids flow, flames burn, and mechanical systems (such as a pendulum) behave. It comes equipped with several interchangeable experiments, called "payloads". A schematic and description of the demonstrator and payloads are given here, followed by suggestions for how you can build your own. In addition, duplicate demonstrators will be distributed to NASA Centers nationwide.

The Reduced-Gravity Demonstrator

The reduced-Gravity demonstrator consists of four essential parts: (1) an experiment platform, (2) a drop structure to support the platform as it is hoisted to the top and released into free-fall, (3) a foam-padded crate to catch the platform as it lands, and (4) a video cassette recorder (VCR) and a monitor to record the experiment during the drop and replay it afterwards.

(1) The experiment platform consists of a video camera and one of several interchangeable experiment payloads (described below). The experiment payloads use base plates of a common size for simple attachment. An eye bolt is fastened to the platform so it can be hoisted to the top of the drop structure using a rope and pulley. The drop is initiated by simply releasing the rope.

(2) The drop structure provides a rigid means by which one can easily raise and release the experiment platform. The structure is segmented for easy assembly and provides a free-fall distance of approximately 2 m. Thus, the platform and experiment will free fall for about 0.6 s.

(3) The foam-padded crate cushions the impact when the platform hits the bottom, and as shown in fig. 1, a mount for the drop structure is fastened to its internal side wall.

(4) The video system consists of a video recorder and monitor that are connected to the video camera mounted on the experiment platform. A long cable is connected from the "video out" port of the camera (on the platform) to the input of the VCR. The cable is approximately 4 m in length to allow unrestricted motion of the platform during the drop. The output of the VCR is then connected to the input of the monitor. It is helpful to use a video recorder with a frame-by-frame job shuttle (or slow motion) to replay the video slowly since the test time is short (0.6 s).

The Payloads

Here are several sample experiments that we use to illustrate the physics of fluids, flames, and mechanical systems. Their behavior is shown both at normal gravity (1g) and also during the drop when the experiment is in microgravity (ug).

Weightlessness. To demonstrate the principle of weightlessness, a scale holding a mass is dropped. The two forces that counterbalance each other in this experiment are the force of the spring in the scale and the gravitational force acting on the mass, i.e., the weight. Prior to the drop, the weight of the mass balances the repulsive force of the spring such that the net acceleration of the payload is zero ($F = ma = kx - mg = 0$, where k is the spring constant and x is the displacement of the spring as indicated by the location of the arrow on the scale). During the drop, the only force acting on the weight, relative to the payload, is the spring force ($\Sigma F = kx = 0 \rightarrow x = 0$), so the indicator on the scale moves toward zero.

The only special material needed for this experiment is a small scale (dietary or postage) with a weight fastened to the top. The weight should be appropriately chosen so that the indicator reads in the middle of the scale before initiation of the drop. We also recommend putting a piece of foam between the weight and the top of the scale to cushion the impact when the package hits the bottom.

Fluid Interface. The fluid-interface experiment highlights the importance of surface tension in the absence of gravity. In 1g, surface tension is evident only near the container walls while most of the surface is flat. In ug, surface tension leads the liquid to assume a very different shape since the liquid tends to creep up the container walls by capillary forces. This is most evident in the corners. Given enough ug time, the liquid would wet all of the walls of the vessel, leaving a bubble of air in the center.

For this experiment, you will need a closed container partially filled with fluid. It is helpful if the container has sharp corners where the effect is most evident. Our acrylic container uses a flat front with a slightly curved back, giving two sharp corners. For the fluid, we recommend a low viscosity silicon oil. Water does not show the gravitational effects well because of its large (relative to silicon oil) contact angle. Other fluids may be used, but their viscosities must be low for the effects of gravity to be evident in the fall time.

Candle Flame. The candle-flame experiment demonstrated the effect of buoyant convection and its absence on combustion phenomena. In 1g, the combustion gases are much hotter, and thus lighter than the surrounding air. Buoyancy causes the hotter, less dense combustion gases to rise, giving the candle flame its vertically elongated, conical shape. However, during the drop, the hot gas expands in all directions since warm air no longer rises. As a result, the flame becomes shorter and wider. In longer periods of µg, the flame becomes spherical and entirely blue. This was observed in a candle flame experiment performed on the Space Shuttle.

This experiment requires a candle with an enclosure to protect it from air drafts during the drop. Otherwise, the candle tends to flicker, obscuring observation of its microgravity behavior. The enclosure must have an opening for air exchange or the candle will quickly extinguish due to depletion of oxygen. Our best success was with a single opening at the top, covered with a fine wire mesh to limit the drafts, and a small door at the base of the enclosure to allow for ignition. This door was closed seconds before the drop; otherwise the flame would extinguish. It may be possible to adapt a small lantern from a camping supply store as the enclosure.

Magnetic Repulsion. Two magnets are oriented with like polarities opposing one another (i.e., N-N pr S-S). The lower magnet is fixed to the experiment platform while the upper magnet is freely supported on a lever arm. In 1g, the upper magnet is levitated by the repulsive force of the lower magnet balanced by the gravitational force pulling the upper magnet downward; i.e., $F_{repulsive} - mg = 0$. During the drop, the only force acting in the falling reference frame is that due to the magnets. Hence, the separation distance between the magnets increases.

This experiment requires two magnets. One should be fastened to the payload base while the other should be suspended above it. We have fastened the upper magnet to a pivoting support, yet there are many different ways to do it. For instance, if you have two doughnut-shaped magnets, you could simply attach one magnet to the base, mount a vertical rod through its center hole, and then suspend the second magnet above the first, using the rod as a guide. A stop should be added at the top of the rod to prevent the magnets from falling off during the drop or upon deceleration. We discovered that ceramic magnets often break upon direct impact. We recommend avoiding them for this experiment or protecting them with an intermediate layer of foam.

We encourage you to devise and try some of your own "drop tower" experiments. These might include studying how and why air bubbles rise in a liquid (demonstrated by placing an antacid tablet in water and dropping it), how fluids flow, how a mass on the end of a rubber band moves, or how the motion of a pendulum changes when it is dropped. For further information on how to incorporate lessons on microgravity into your curriculum, we encourage you to get a copy of NASA's brochure describing the Reduced-Gravity Demonstrator along with the teachers' guide. These are available on the World Wide Web (WWW) at http://zeta.lerc.nasa.gov or from your nearest NASA Teacher Resources Center. The guide briefly discusses the history of gravity and the relationship between weightlessness and free fall, then describes a number of other experiments pertaining to gravity that can be easily demonstrated to students.

12

Gaint Newton's Rings

The classic demonstration of Newton's rings (the pattern of concentric colored circles seen when thin films are bound together) is a popular one in introductory physics courses. Usually a plano-convex lens of large radius is placed on a sheet of plane plate glass and light from a monochromatic source falls on the arrangement at right angles to its surface. When the light is reflected into an observer's eye, the point of contact of the lens appears as a dark spot surrounded by a series of circles. These circular "interference fringes" are produced by the circular symmetry of the air film sandwiched between the curved surface of the lens and the upper surface of the glass plate.

Unfortunately the rings produced in this manner are only barely observable to the unaided eye. For instance, with a lens of R = 1 m, the radius of the fringe pattern is of the order of 1 mm. Observation and measurement are usually done with the help of a traveling microscope mounted directly above the fringes.

We describe here a simple modification of the

traditional setup that magnifies the scale of the interference pattern so the demonstration can be used for the whole class or for semiquantitative measurements in any high-school laboratory. A set of fairly clear Newton's rings of up to 2 cm in diameter can be produced without need for precision instruments. Also, not only are the fringe patterns observable from almost any direction above the glass plate, but their size and shape are independent of the position of the observer.

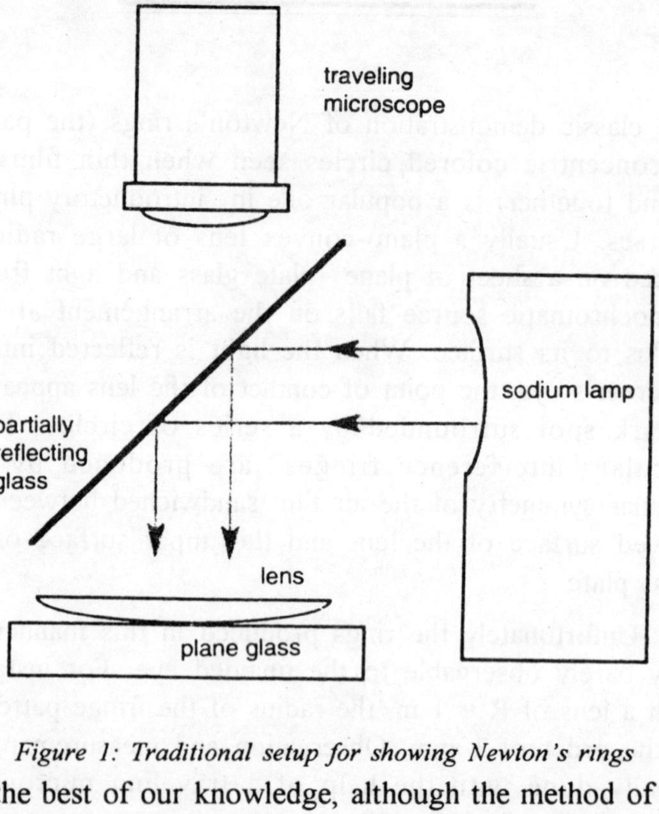

Figure 1: Traditional setup for showing Newton's rings

To the best of our knowledge, although the method of the formation of giant Newton's rings was published in Chinese in 1988, this column contains the first discussion of the theory involved.

Arrangement

Support a piece of white and smooth Plexiglas (about 30 x 30 cm, 3 mm thick) on a coin at its center of mass. Bend the Plexiglas slightly convex upward by placing 1-kg standard weights at the four corners. Place a smaller piece of glass on top of the Plexiglas at its center point. Direct white or filtered colored light (from a 100-W bulb or overhead projector) onto the plates at near-grazing angles. With a partial blackout, students from all directions can observe Newton's rings above the plates of glass.

Theory and Sample Experimental Result

The choice of a suitable white Plexiglas plate and a large angle of incidence for the incoming light are two crucial factors leading to a successful demonstration.

Like most opaque dielectrics, the white Plexiglas produces partial regular reflection on the smooth surface but diffuse reflection by impurities inside the material. From Maxwell's equations, or by a simple experiment, we can show that the percentage of partially and regularly reflected light to the incident light increases as the angle of incidence is increased. For a small angle of incidence, i, most of the incoming light passes through the surface and suffers diffuse reflection. However, the percentage of partially reflected light increases drastically when $i > 70°$ and becomes 100% when i approaches 90°.

The shape and the size of these circular fringes, like that of a real image caught on a while screen in geometrical optics experiments, are independent of the position of the observer. This suggests that the interference pattern is first formed on the upper surface of the white Plexiglas plate and then reaches the eye through diffuse reflection. This conjecture is further supported by an ad hoc change of the arrangement. When the white Plexiglas is replaced with clear Plexiglas, we find that the giant fringes almost completely disappear. However, if we place a piece of white paper under the Plexiglas plate, the fringes reappear on the white paper. The image gets blurred and finally disappears as we move the paper down to a lower position. No doubt these fringes are formed by transmitted light rays through the air film and caught on the white paper that acts as a projection screen. For this reason, we can consider the white Plexiglas plate as a combination of clear Plexiglas and a white screen.

Radii of giant Newton's rings

	measured r	*calculated r*
1st bright ring	7 mm	6.6 mm
1st dark ring	8.5 mm	8.1 mm
2nd bright ring	10 mm	9.3 mm
2nd dark ring	11 mm	10.5 mm
3rd bright ring	12 mm	11.5 mm

Consider two light rays, 1 and 2, traveling from a point on the light source to a point P on the upper surface of the white Plexiglas plate. Since the lower Plexiglas plate is bent by force, the variation in thickness of the upper glass plate is much smaller than that of the air film. Under this approximation, the glass plate introduces the same lateral displacement to both light rays and is used solely to provide a boundary to split the amplitude of the incoming light.

Although ray 2 suffers two more regular reflections (at A and B) than ray 1, they are still of comparable amplitudes because the angles of incidence of the incoming rays are close to 90°. The rays that follow the paths shown are seen by the eye to interfere at P.

The path difference between these two light rays is

$$\frac{n\lambda R}{\cos\phi}$$

Since reflections at both A and B occur on the air-to-glass boundary, there will be no phase change in these reflections: a white spot will be observed at the point of contact. The conditions for constructive interference are simply

$n\lambda = 2d \cos \phi,$ $n =$ o, 1, 2.............

From the sagitta theorem, let r be the radius of a Newton's ring, $(2R - d)\, d = r^2$.

For $d << R$,

$$r^2 = 2Rd$$

From the path-difference formula, the radius of the nth bright ring is

$$(r_n)^2 = \qquad (3a)$$

and the radius of the (n + 1)the dark ring is

$$(r_n')^2 = \frac{(2n+1)\lambda R}{2 \cos \phi} \qquad (3b)$$

By using an ordinary meter rule and a thickness gauge, we find that when $d = 0.3$ mm, $r = 6.5 \pm 0.5$ cm. Substituting these values into Eq. (2), the radius of curvature of Plexiglas plate is $R = 6.5 \pm 0.6$ m.

In a typical experiment, we use filtered red light with $\lambda = 0.7 \times 10^{-6}$ m derived from and overhead projector. The angle of incidence is fixed at = 84 °. The radii of rings are measured directly and calculated using Eq.(3). The measured and calculated results are in good agreement with one another with an understanding that an error of slightly over 10% is assumed in each column.

13

A Simple Electronic Circuit to Demonstrate Bifurcation and Chaos

Chaos has generated much interest recently, and many of the important features of nonlinear dynamical systems can be demonstrated using the simple electronic circuit described here.

The study of chaos has generated enormous interest in exploring the complexity of behaviour shown by nonlinear circuits and other dynamical systems . A simple electronic circuit, which could form an introduction to project work in the field of chaos, is Chua's autonomous circuit.

Chua's circuit can be used to demonstrate many of the important features of nonlinear dynamical systems, such as bifurcation, period doubling, attractors and chaos. We present here a simple description of the operation of this circuit, a practical electronic implementation, and discuss some of the experimental observations that can be made with it.

Chua's circuit

The basic circuit, is reciprocal and consists of only one

nonlinear elements. The parallel combination of C_2 and L constitutes a lossless resonant circuit. The conductance G provides the coupling between this, the active nonlinear resistor R_1 and $C1$.

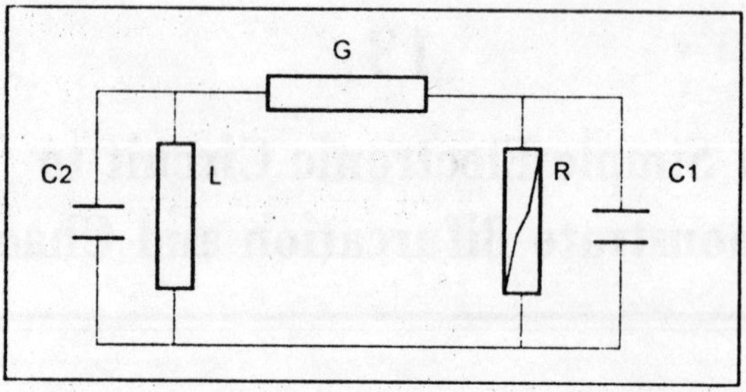

Schematic diagram of Chua's basic circuit

Since R is active, it behaves as a power source. The idealized nonlinear element R is a three-segment piecewise-linear resistor with the transfer characteristics.

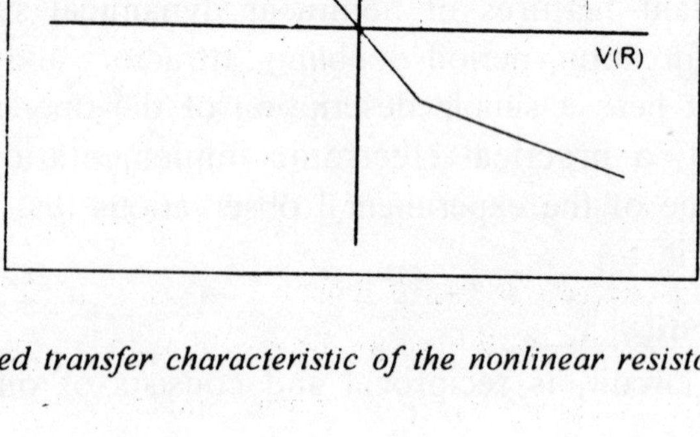

Idealised transfer characteristic of the nonlinear resistor of R

The circuit dynamics is described by

$$C_1 \frac{dv_{C_1}}{dt} = G(v_{C_2} - v_{C_1}) - g(v_{C_1})$$

$$C_2 \frac{dv_{C_2}}{dt} = G(v_{C_1} - v_{C_2}) + i_L$$

$$L \frac{di_L}{dt} = -v_{C_2}$$

where V_{c1}, V_{c2} and I_L denote the voltages and currents acorss and through the passive components, and g denotes the transfer function of the nonlinear resistor.

We refer the interested reader to the technical papers by Matsumoto *et al* for a detailed investigation of these equations.

Practical implementation

The only non-trivial circuit element to implement is the nonlinear resistor R. There are several techniques that have been used in practical circuits. It can be synthesized by a single op-amp, two diodes and resistors, two bipolar transistors, two diodes and resistors, or two op-amps and six resistors. We have found that the simplest circuit to implement in practice is that of Kennedy using two identical op-amps. This circuit, has been used in our undergraduate teaching laboratories for demonstration purposes. The choice of op-amp is not critical, the tolerance on the values of the capacitors and the inductor is $\pm$ 10% and the fixed resistors' tolerance should be $\pm$ 5% or better. In an experiment the value of the conductance, implemented as a series pair of two non-inductive variable resistors, is slowly changed.

The *X* and *Y* channels of an oscilloscope are connected to the non-earthed ends of C_1 and C_2. By this means one can observe the change from a period-one can observe the change from a period-one attractor, via a sequence of period doubling bifurcations, to a spiral attractor, periodic windows and a double-scroll chaotic attractor.

A qualitative description of the behaviour demonstrated by Chua's circuit

The role of the Control parameter in this circuit is taken by the conductance G. The active, or driving, part of the circuit has a nature frequency, w, associated with it which depends on the specific reactance of its components. In addition, the *I-V* characteristic of the active, forcing part of the circuit is symmetric under a rotation of 180 degrees, or the current through, the passive part of the circuit may be measured as a function of time. Alternatively, both these quantities can be displayed simultaneously along the *x-y axes* of an oscilloscope.

When the apparatus is switched on, initially, one observes a *transient,* which settles to a *steady state* response. This steady state response is called an attractor of the system. This means that the responding, or passive, part of the system can take many different initial conditions which will all lead to the same steady state behaviour. For small or large values of the conductance we observe that the values of both the current and the voltage in the passive part of the circuit return to the same values periodically. In fact, they repeat with the same frequency, ω, as that of the driving force, and thus we say that we observe a *period-one attractor*. As the conductance is smoothly increased we observe a discontinuous change in the response, or a *bifurcation.* The period-one solution period doubles and thus the period of time over which the motion repeats is twice as long.

Period doubling is a phenomenon that is considered to be universal in nonlinear dynamical systems. The same behaviour is observed in many physical systems, for example the periodically forced simple pendulum and mathematical models, for example predator-prey *dynamics*. *A period doubling cascade* is a common route to the observation of chaos. Here, in Chua's circuit, as we smoothly increase the conductance we see the period-2 period double to a period-4, similarly, the period-4 period doubles to a period-8. With infinite precision, we would expect to see an endless series of period-doublings as successively smaller and smaller increases in the control are made, with the limit being a cycle of infinite period, displaying all the characteristics of a *chaotic attractor*.

In practice, we have found that we can easily observe the period-2, period-4 and period-8. A further small change yields a single scroll chaotic attractor, its shape characteristic of Chua's circuit. An interesting, and counterintuitive, property of chaos is that coexisting with the chaos are an infinite number of periodic solutions of all periods. Careful tuning of the control parameter may result in the abrupt disappearance of the chaotic attractor, which is replaced, for a small interval, by a periodic solution. The chaotic attractor will suddenly reappear at the end of this interval. This phenomenon is known as the appearance of a *periodic window*. In Chua's circuit, we can observe a period-3 window, which itself period doubles, and three separate period-5 windows. The observation of these windows in itself is evidence that we have a chaotic solution rather than noise. As mentioned above, the nonlinear restoring force is symmetric under a rotation, about the origin, of 180 degrees. This means that two periodic attractors are available to the system. Which one the initial transient eventually settles to depends on the initial condition. As our control is increased we observe one final bifurcation. The attractor suddenly spreads to fill the entire region previously occupied by the two smaller attractors, producing the *double scroll chaotic attractor*.

14

Measurement of Radioactivity in Buildings with a Large-Area Silicon Detector

Natural background radiation is mainly due to radon, but how much is in the air around us? A figure can be calculated by using a silicon detector to measure — particles emitted from the dust deposited on a filter through which air is drawn by a vacuum cleaner. The detector described here also allows measurement of very low macroscopic surface activities.

The air in our schools and homes contains trace amounts of radioactive radon. There are three common radon isotopes, ^{219}Rn, ^{220}Rn and ^{222}Rn, formed in decay chains of ^{235}U, ^{232}Th and ^{238}U nuclei, respectively. Of these three radon isotopes most of the radiation comes from ^{222}Rn, which has a half-life of 3.8 days. The main source of this isotope is soil. Radon, a chemically inert gas, seeps up through foundations of building, becomes trapped in the interior and then decays. The radon daughter nuclei ^{218}Po, ^{214}pb and ^{214}Po are chemically active and can deposit in the lungs when inhaled. Inside buildings the radon daughters adhere to dust particles, walls or other macroscopic surfaces. The average radon indoor concentration ranges in most European countries from 20 to 50 Bq m^{-3}, while in Scandinavia the average is about 100 Bq m^{-3}.

The presence of radon daughters adhering to dust particles can be detected in the atmosphere by pumping air through a filter, and then placing the filter close to a suitable detector for measuring the radioactive contamination. Such an experiment was performed by Lumb, who used the Geiger-Muller tube was a detector of α and β particles. He observed a decrease in time of the activity due to radon daughters.

Activity concentrations (μBqm⁻³) of natural radionuclides in air.

^{238}U-^{234}U	1
^{230}Th	0.5
^{226}Ra	0.5
^{210}Pb	50
^{232}Th	1
^{238}Ra	1
^{238}Th	1
^{235}U	0.05

Silicon detector

In the present work we registered radon prog— means of a large silicon epitaxial detector with an active area of about 64 sensitive to α particles. The p-n junction was produced in an epitaxial process at the Warsaw Institute of Electronic Materials Technology on an n^+ silicon wafer substrate 10 cm in diameter and 400 μm thick. To extend the depletion layer, which is depleted of free electrons and holes by an electric field, the detector p-n junction is polarized by applying reverse bias via a bias resistor. The α particles passing through the detector produce electron-hole pairs which are immediately separated in the junction by an electric field. Holes move to the p-type silicon where a negative potential is applied, while electrons move in the opposite direction to the n^+-type material. The total charge generated in the detector by α particles is registered by a charge preamplifier followed by additional electronics.

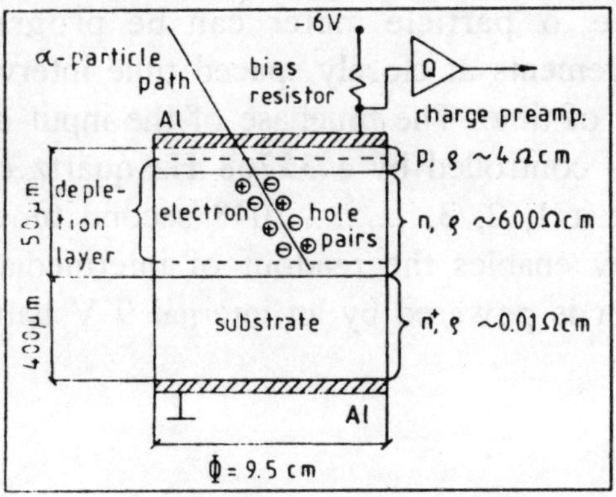

Figure: Epitaxal silicon detector

Since the amount of charge generated is proportional to the kinetic energy lost buy the particle in the detector, we can calibrate the detector by using - particle sources of known energy.

The - α particle meter APM-45

For measurements of radon daughters we designed the α particle meter the block diagram of which is shown in the following figure. The use of the charge-sensitive preamplifier as the first stage is a natural choice since the detector has a very large capacitance. It makes the amplitudes of the output signals from the preamplifier relatively independent of the detector capacitance. The following one-stage *RC-CR* shaping amplifier converts the signals from the charge preamplifier to about 10 μ wide gaussian pulses. Such noise filtering enabled us to set a 1 MeV threshold in the following pulse height discriminator. Pulses with amplitudes above1Mev trigger the discriminator, which delivers standard logic pulses that are stored by the following counter. The gate generator provides start, stop and reset pulses controlling the counter. The α particle meter can be programmed to take measurements at closely spaced time intervals over a long period of time. The timebase of the input of the gate generator is controlled by a 32768 Hz quartz clock. One may choose a 1, 2, 3, ...2048 second timebase. The LCD display enables the readout of intermediate results. The α meter is powered by an internal 9 V battery.

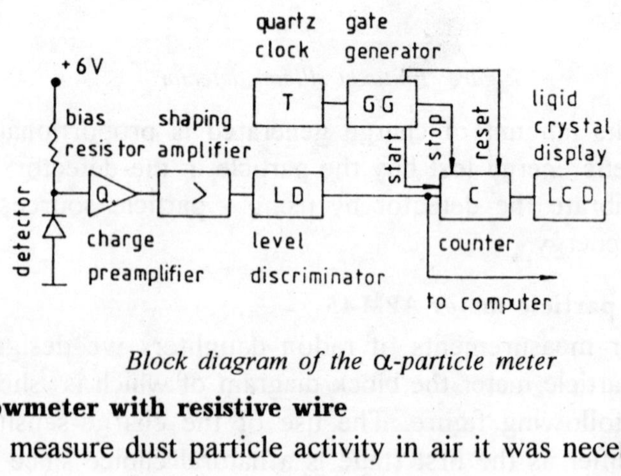

Block diagram of the α-particle meter

Flowmeter with resistive wire

To measure dust particle activity in air it was necessary to measure the volume of air pumped through the filter by the vacuum cleaner. We employed a simple flowmeter based on the cooling of an electrically heated resistive wire in the air stream. The temperature of the wire was kept constant at fixed value in order to satisfy the condition of heat transfer between wire and air. If the airflow was increased more electrical power had to be supplied to keep the wire temperature at the fixed value. Thus more heat was dissapated in the wire to balance the increased heat transfer to the air.

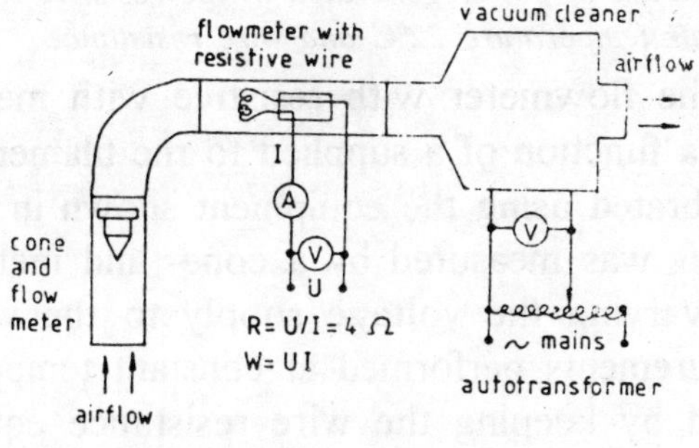

Diagram of the flowmeter with resistive wire and equipment for calibration

The design of the flowmeter is shown in figure, The flowmeter consists of a plastic pipe and a resistive wire mounted inside. We used a filament from the 6.3 V/0.3 A incandescent lamp. The resistance of the cold (22°C) filament is about 2Ω. When the lamp was nominally powered the temperature of the filament rose above 1000°C and the resistance approached 21 Ω. the above observed increase of the filament temperature as a function of its resistance is taken advantage of in our measurements. To stabilize the filament temperature we kept the filament resistance constant. After smashing the glass bulb of the lamp we mounted the filament inside the pipe.

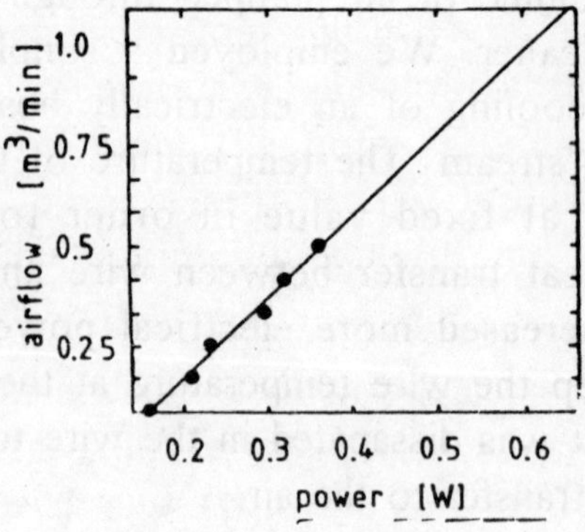

Airflow as a function of power generated in the resistive wire, for air temperature 22⁰C and wire resistance

To calibrate the flowmeter with resistive with measured the airflow as a function of a supplied to the filament. The flowmeter calibrated using the equipment shown in figure. The flow of air was measured by a cone- and meter and regulated by varying the voltage supply to the vacuum cleaner. Measurements performed at constant temperature of the ensured by keeping the wire resistance equal to. The air temperature was 22⁰C. The calibrate procedure consisted of the following steps;

1. Adjustment of the airflow by setting the voltage supplied to the vacuum cleaner the help of an autotransformer.

2. Adjustment of the voltage U across the W at which the wire resistance R reached 4 i. e. $R= U/I =4\Omega$ where I is the measure of electric current passing through the wire.

3. Calculation of the power W delivered to

$$\text{wire} : W = UI$$

The above three step were performed for six values of airflow measured by the cone-and-float meter. Figure presents a diagram of airflow versus power delivered to the wire. The dependence is linear.

Measurement of air activity due to radioactive dust particles

For this measurement we produced a source of - particles derived from and environmental; air sample. The sample is generally expected to be of very low specific activity and it is necessary to collect to large volume for analysis. A know volume of air was drawn through the filter, and then the filter was measured by means of an - particle meter. The sample activity A is obtained from the formula where *N* is the number of counts made during time t [s], ε is the α counting efficiency of the detector determined by the solid angle. (about 45%, i.e.ε = 0.45). The symbol η is the absorption factor of the filter and (for our petrianov filter $\eta = 0.98$) and V[m^3] is the volume of the air sample drawn through the filter. In figure 6 the set-up for collection of dust is illustrated. The filter was supported on a mesh and mounted in the dust collector. The air pumped by a vacuum cleaner passed through the filter, which was placed 5 mm from the silicon detector. The flowmeter with resistive wire was placed between the vacuum cleaner and the filter. After t = 128 second of collecting data N = 25 α particles were recorded. The volume of air drawn through the filter was measured by the flowmeter with resistive wire. An airflow of 0.75 m^3 min^{-1} was obtained. After 128 s of pumping a sample of air V = 1.6 m^3 had passed through the filter. The activity $A = 0.28 \pm 0.05$ Bq m^{-3} was evaluated from formula. Since the measurement and collection time of radon daughters was short compared with the time scale of fluctuation in activity concentration, the method of measurement presented here was classified as a grab sampling method.

Decay of radioactivity in dust

A piece of toilet paper of diameter 12 cm was placed in the dust collector as a filter. The dust was collected on the filter and its activity was recorded. The measurements were made by an IBMPC computer to which the standard pulses generated by - particles in the α particles meter were transferred. A DCI Smartlab 8255/8253 CARD was used. The computer was programmed to work in the multiscaler mode. Each measurement lasted 3 minutes. Air was pumped through the toilet paper for 15 minute and at the same time the computer was collecting data. On the computer monitor the number of counts is displayed as a function of time. The activity increased during pumping and decreased after pumping was stopped. The obtained decay half-life was about 45 minutes. this is the total effect of the decays of ^{212}pb ($T_{1/2}$ = 26.8 min) and ^{214}Bi ($T_{1/2}$ = 9.7 min) which are in the same radioactive chain. A cable connected to the α particle meter was used for the power to the device and for transfer to the computer standard logic pulses generated by- particles. Air was delivered to the filter through a circular between the α particle meter and the dust collect.

15

Development of Multiple-Choice Test Items

The "good" multiplc-choice test is an elusive entity. There seems to be an understanding in the physics community that multiple-choice tests give at best an approximate idea of the student's learning, and at worst an artificial score based on random guesses on a multiple-choice exam with four choices per item, it is possible for that student to get (on average) 25% of the questions correct without implementing any thought. Sometimes teachers try to prevent random guessing by deducting one-quarter of the value of a question one-quarter of the value of a question for each wrong answer. Higbie suggests offering an "I don't know" option on each question. The student who opts for the "I don't know" answer is not penalized, whereas a penalty is charged for incorrect choices. This allows the instructor to assign a measure of uncertainty in the score, which can be a separate measure from the score itself. Higbie adds, "If the score is within one standard deviation of a grade boundary, it should be promoted over that boundary." This gives the student the benefit of the error, a strategy with which I tend to agree.

Having our students work out problems on exams can tell us the most about the thought processes used, and if the number of students to be evaluated is small, or there is ample staff for grading papers, then such problem solving is arguably the preferred form of evaluation. However, in the evaluation of large numbers of students, the multiple-choice exam is the only viable option. If properly constructed, the multiple-choice exam can also offer insight into the thought processes of the physics student. As mentioned, certain strategies help in the calculation of an accurate score, but not test can be better than its questions, so some consideration needs to be placed on the development of the individual questions.

Some test evaluation basics

The key process in the evaluation of a test question is item analysis, which broadly can be viewed as a process of comparing the individual item to the entire test. (We have to assume that the exam as a whole is approaching what we would like to test for; the process of item analysis can be viewed as a refinement process.) Most usually the evaluation is done on a computer system that has an optical scanning device to read the students' marks on the answer sheets. Typical item-analysis printouts show the number or the percentage of students who selected each response, as well as the average score of the students who picked each item. There is also some number that relates the individual item to the test as a whole. This can be called the discrimination index, or can be expressed as a correlation coefficient. The discrimination index is defined as $D = \frac{H-L}{N/2}$ where N is the total number of students, H is the number of students who got the question correct and are in the top half of the test scores, and L is the number of students who got the question correct and are in the bottom half of the test scores. A question with a very high discrimination (+1.00) will set the high and low students apart perfectly. A question with a discrimination of 0.00 cannot tell them apart at all. A question with a low discrimination (-1.00) would be a had item, since the students who did poorly on the test over-all are doing better on this question.

The measure we use in the IUP (Indiana University of Pennsylvania) Physics Testing Competition is the Person product-moment correlation coefficient:

$$r = \frac{\Sigma XY - (\Sigma X \Sigma Y)/N}{\Sigma X^2 - [(\Sigma X)^2/N]\,[\Sigma Y^2 - (\Sigma Y)^2/N]} \quad (1)$$

Here, the N represents the total number of students. Y the score for the particular question, and X the score of the test overall. IF the correlation is near the value of positive one(1), then students performed as well on this question as they did overall. The closer the value is to one (1), the better this question is representative of the exam. If the correlation is near the value of minus one(-1), then the students' performance on this question is inversely representative of their overall score. The closer the value is to minus one, the worse this question is representative of the exam.

If we assume that the test on average is a good test, we can then inspect the questions that have a low correlation coefficient and adjust them. If it seems that the wording is confusing or the question is misleading, we can change the question. It is also interesting to observe the average scores of the students who selected the incorrect distracters. If they are as high as the scores of the students who selected the correct answer, then that question to requires some investigation.

Once a test has been developed that adequately measures the physics knowledge of students, it can be used to measure the relative success of physics programs. To compare students from various schools, in 1993 we set up a competition based on a 100-question multiple-choice exam. The first year there were 1413 students representing 63 schools in Pennsylvania. The exam is taken during April and the answer pages are sent back to IUP for scoring. Top students and the top schools are recognized for their achievement. Students who have taken physics for either one year (7 months) or two (14 months) are eligible to take the same test; they are recognized in separate categories. Item analysis is done separately for the two groups, as well as for the whole group. Three specific items from the 1993 test are discussed here.

This question had the lowest (r = 0.00) correlation coefficient:

> Newton's Second Law states that the acceleration of an object of given mass is proportional to the force that is applied. We also know that a 20-kg object by gravity is twice as much as that of the 10-kg object. Since the force on the 20-kg object is twice as that of the 10-kg object, why does it not accelerate twice as fast as stated by Newton's Second Law? The reason for this can be explained by which of the following?

A) its weight B) its mass C) its inertia

7% 23% 32%

28.6		30.31	29.41	
D) A & B	E) B & C		Omit	
9%		33%		8%
25.59	27.97		21.29	

The first row of numbers is the percentage of the students who selected that response. The second row of numbers represents the average score of the students who selected that response. The correct answer (E) was selected by only 33% of those who responded. The average score for the students who selected that answer was 27.97. Almost as many students (32%) selected choice C, and the students who selected that answer had a slightly higher average score. It is possible that the wordiness of the question distracted some of the students: they may have forgotten what the question was about by the time they finished reading the question. Also, it is possible that some did not realize the relationship between inertia and mass. The better student may realize that mass and inertia are different, reads the question, and picks the first answer that comes to mind, without realizing that objects with more mass also have higher inertia.

Granted, students should be able to read, and draw their own illustrations from a given text. However, the item analysis indicates that something is amiss with this problem, because it confused even the better students. This question performed a bit better with students who had up to 14 months of physics; the correlation coefficient was 0.19.

The following question had the second lowest correlation coefficient (r = 0.07)

Distance is to displacement as

A)	mass is to weight	31%	29.34
B)	work is to energy	25%	28.12
C)	heat is to temperature	22%	29.80
D)	acceleration is to force	12%	24.90
E)	electricity is to magnetism	14%	24.16
	Omit	6%	21.34

In this question the relationship of a scalar quantity to a cognate vector quantity is being explored. The student who is very skilled at physics would probably realize that distance is a scalar quantity and that displacement is a vector quantity, but would not be drawn to the accepted answer, considering that mass and weight are not related in quite the same manner since there is a constant of proportionality. The correct answer (A) was selected most often (31%) and the average score for the students who selected this answer was 29.34. The average score of the students who selected C had the highest score on the test, 29.80. In general, it seems that the question does not work well since the relationship between mass and weight is not quite the same as that between distance and displacement. A more appropriate correct answer for this question might be "speed is to velocity". This question also had a slightly higher correlation coefficient ($r = 0.16$) for the students who had physics for 14 months.

These two questions were deemed as least related to the test as a whole. Here is the question with the highest correlation coefficient ($r = 0.48$).

The frequency of the wave is:

A) 66 B) 99 C) 163 D) 198

9%	7%	31%	36%
26.70	26.67	27.15	36.40
E) 396	Omit		
4%	33%		
27.07	19.75		

The reason for the high correlation of this question can be seen from the larger average score for students who selected the correct answer (D). This question was part of a four-question set where the student is asked to find the velocity, wavelength, period, and frequency of the wave. These four problems had the highest correlation on the test for the students who had 7 months of physics. The possible success of this question is that it requires a multiple-step process: first determining the velocity of the wave, then finding the wavelength of the wave, and finally using the relationship between the wavelength and the speed of a wave to find the frequency. It can be difficult for students to consider more than one step in solving a problem.

16

Study of Three-Dimensional Images of Sources Using Biprism

Introduction

In literature, Euclid's principle of geometry reveals the perception of depth from flat paintings and three-dimensional reality in picture. Studies on the pictorial memory, visual concepts of reading and the role of visual processing in perceiving the layout of space have been done by many authors.

The subsequent development of this scientific field appeared practically in 3D movies with three spatial dimensions, length, breadth and depth. The prime requirement of 3D vision is two viewing points separated by a distance ee' and the interaxial angle α. The two images transmitted by eyes are to be processed and interpreted by the brain of the viewers to give a single composite 3D image. Therefore 3D vision may be described as a perception of depth and distance based on binocular vision combined with non-stereoscopic cues and possibly other factors not yet fully understood.

The four systems developed for achieving the

perception of depth are (1) *anaglyphic* 3D, (2) *polaroid* 3D, (3) *stereographic* 3D, and (4) *lenticular* 3D. The first uses two colours for the two stereographic images and transmits them through corresponding filters to the viewer for 3D viewing. The second uses two mutually perpendicular polarizations in recording and corresponding viewing through polaroids. The third system uses rapidly alternating black-and-white images and viewing through correspondingly rotating shutters by the two eyes. The fourth uses a lenticular screen which has prismatic vertical wedges. The two stereoscopic views are projected by two projectors on this screen, whose wedges reflect them to left and right eye corresponding to the original recording.

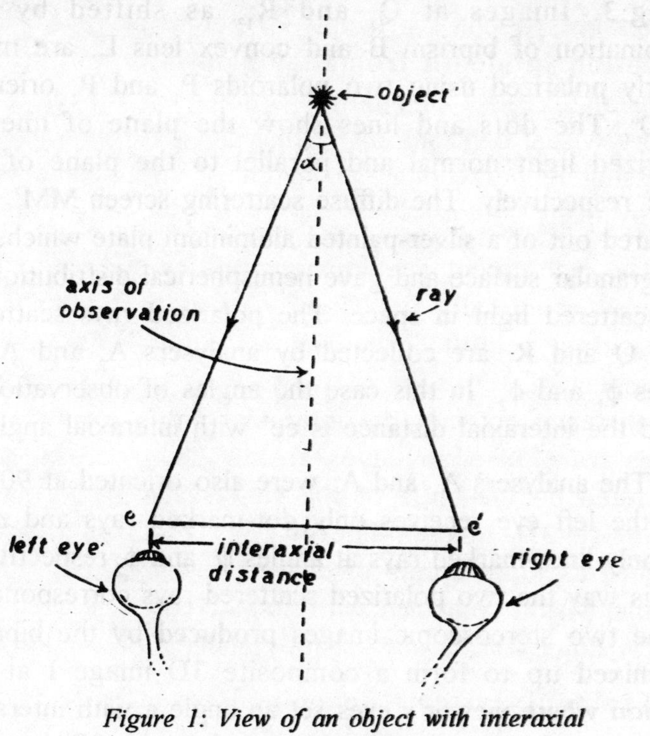

Figure 1: View of an object with interaxial distance ee' and angle α

In this report a simple way of producing the perception of depth for the 3D image is presented using a biprism and a pair of polaroid sheets. Here the purpose of the biprism is to produce two stereoscopic images and to serve as 'two eyes' viewing a field from two angles. Therefore it provides the two shifted patches Q_1Q_2. R1R2 (corresponding to two virtual sources S_2 and S_1 in the line of the source S behind biprism) on a diffuse scattering screen MM' as shown in fig.2.

Now to channelize the two stereoscopic images S_1 and S_2 distinctly into the observer's eyes, two polaroids P_1, P_2, two analysers A_1 A_2, and one converging convex lens L are used as shown in schematic sketch of the setup in fig.3. Images at Q_1 and R_1, as shifted by the combination of biprism B and convex lens L, are made linearly polarized using two polaroids P_1 and P_2 oriented at 90°. The dots and lines show the plane of linearly polarized light normal and parallel to the plane of the paper respectively. The diffuse scattering screen MM' was prepared out of a silver-painted aluminium plate which had fine granular surface and gave hemispherical distribution of the scattered light in space. The polarized rays scattered from Q_1 and R_1 are collected by analysers A_2 and A_1 at angles ϕ_1 and ϕ_2. In this case the angles of observation is 0 and the interaxial distance is ee' with interaxial angle a.

The analysers A_1 and A_2 were also oriented at 90° so that the left eye receives only dot-marked rays and right eye only lines-marked rays at angles ϕ_1 and ϕ_2 respectively. In this way the two polarized scattered rays corresponding to the two stereoscopic images produced by the biprism are mixed up to form a composite 3D image I at the position where viewer's eyes set an angle a with interaxial distance ee'. Now one is able to view the 3D image I

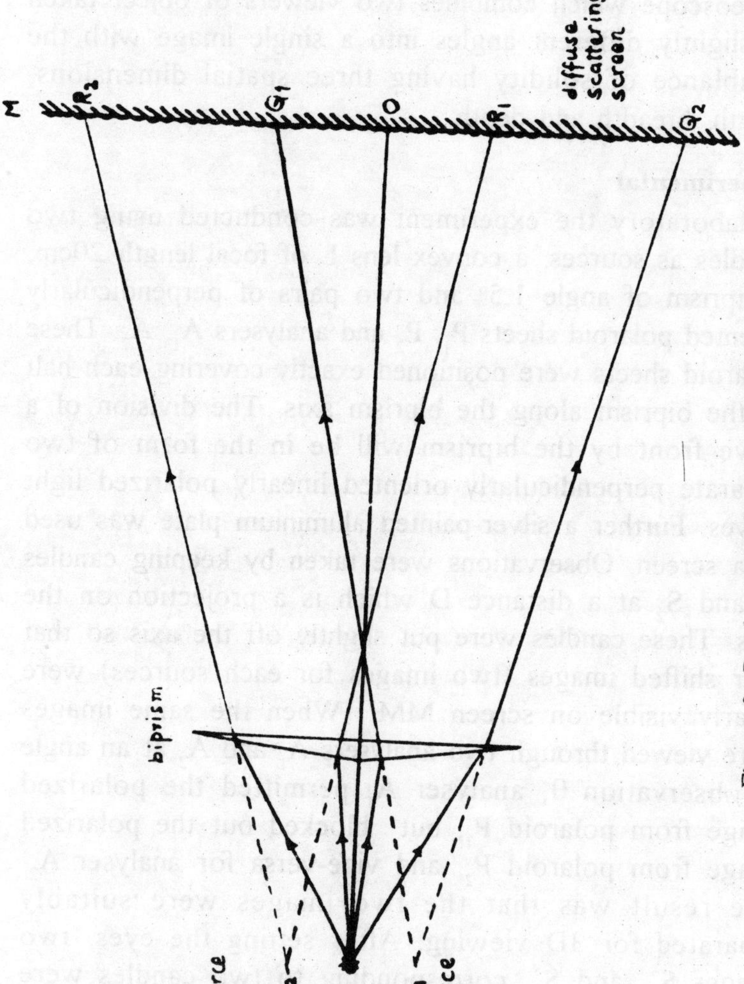

Figure 2: Production of relative shift in patches of two virtual sources S_1, S_2

which appears to be hanging in the air in front of MM'. This gives depth perception and the system acts like a stereoscope which combines two viewers of object taken at slightly different angles into a single image with the semblance of solidity having three spatial dimensions, length, breadth and depth.

Experimental

In laboratory the experiment was conducted using two candles as sources, a convex lens L of focal length 20cm, a biprism of angle 1.5° and two pairs of perpendicularly oriented polaroid sheets P_1, P_2 and analysers A_1, A_2. These polaroid sheets were positioned exactly covering each half of the biprism along the biprism axis. The division of a wave front by the biprism will be in the form of two separate perpendicularly oriented linearly polarized light waves. Further a silver-painted aluminium plate was used as a screen. Observations were taken by keeping candles S_1 and S_2 at a distance D which is a projection on the axis. These candles were put slightly off the axis so that four shifted images (two images for each sources) were clearly visible on screen MM'. When the same images were viewed through two analysers A_1 and A_2 at an angle of observation θ, analyser A_1 permitted the polarized image from polaroid P_1, but blocked out the polarized image from polaroid P_2, and vice-versa for analyser A_2. The result was that the two images were suitably separated for 3D viewing. After setting the eyes, two images S'_1 and S'_2 corresponding to two candles were observed with depth d (projection along axis).

Experimentally the distance *d* between the 3D images S'_1 and S'_2 was measured at one of the observation angles θ to investigate the correlation between *d* and *D*. The observations were made with pointers of the upright on an optical bench using a vernier of the least count of 0.01cm. An uncertainty of the order of $\pm$2mm due to finite diameter of candles was observed in each observation. The measured distance d was plotted as a function of D at constant angle of observation $\theta = 45^\circ$ as shown in fig.5. A linear dependence was found between d and D. Further to test the dependence of *d* on θ, observations were repeated keeping D constant while θ was varied between 0° and 80°. Data are give in Table 1.

Table. 1 Tabulation for depth d at different q. D is kept constant = 6cm.

Observation angle θ (degree)	Depth *d* of images (cm)
0	8.60
10	8.30
20	7.90
30	7.35
40	6.55
50	5.30
60	4.20
70	2.85
80	1.50

Another interesting result was observe when analysers A_1 and A_2 were interchanged. The composite 3D image appears to go behind the screen and recedes away from the observer which is just opposite to the result as obtained in fig.3. In this case a is small due to mixing of rays after extrapolating them behind the screen MM'. This results into negative depth and virtual image I. Hence the measurement of depth *d* is difficult.

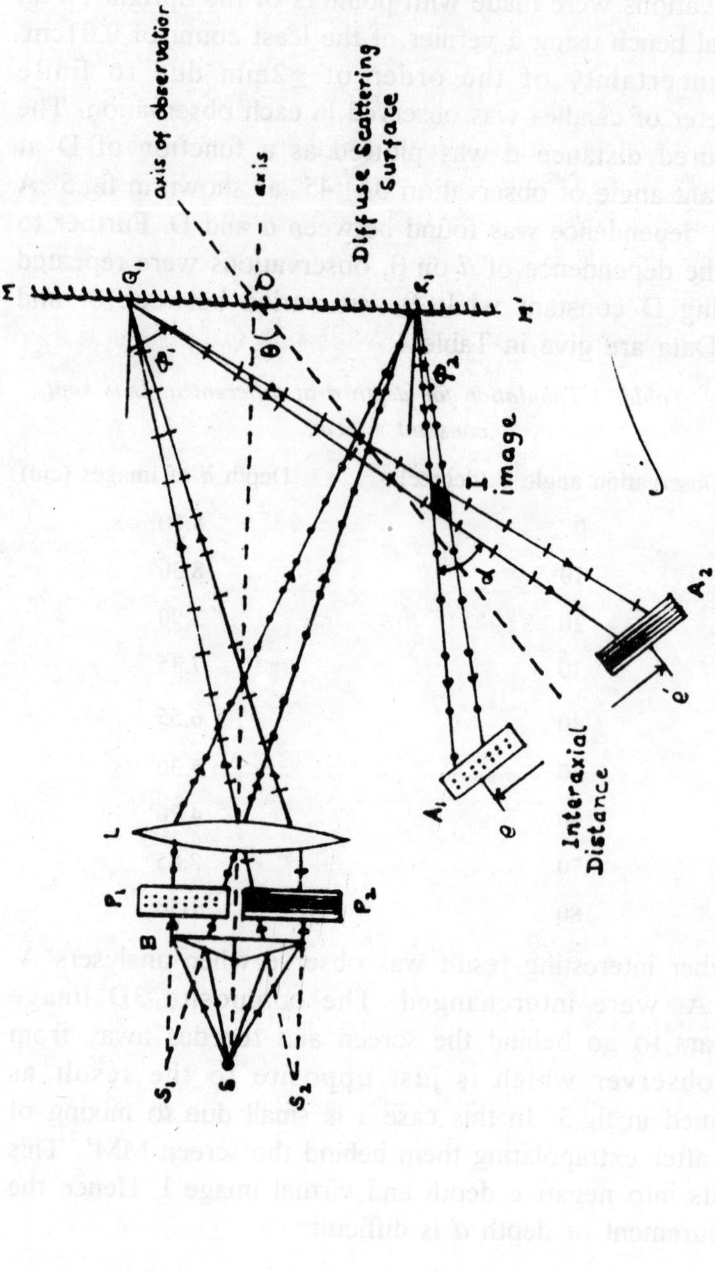

Schematic representation of 3D image of source with optical components

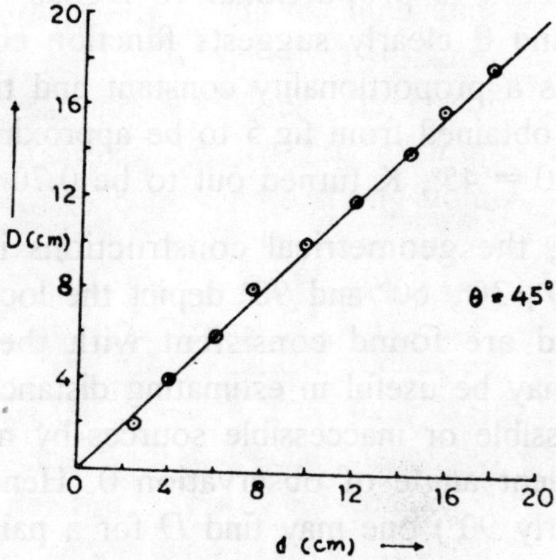

Graph between D and d for θ = 45°

Discussion

The data of d as a function of θ and D were examined and found to exhibit an emperical relation

$$D \cos\theta = Kd$$

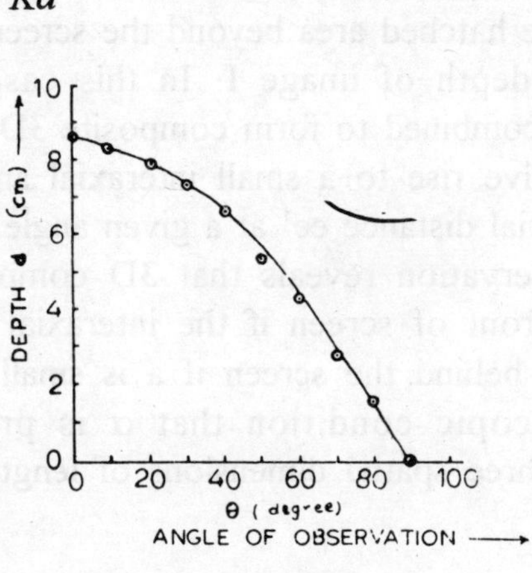

Dependence of depth d of image on the angle of observation θ

It shows that d is proportional to $D\cos\theta$. The graph between d and θ clearly suggests function $\cos\theta$. K was introduced as a proportionality constant and the value of K cos0 was obtained from fig.5 to be approximately 1.00. Substituting $\theta = 45^\circ$, K turned out to be 0.70.

Further, the geometrical constructions in fig.8 for angles $\theta = 0^\circ, 30^\circ, 60^\circ$ and 90° depict the location of 3d image I, and are found consistent with the emperical relation. It may be useful in estimating distance D of any pair of accessible or inaccessible sources by measuring d at a convenient angle of observation θ. Hence at large angle θ (nearly 90°) one may find D for a pair of distant inaccessible source.

After interchange of analysers A_1 and A_2 without disturbing any other arrangement, the 3D image appears go behind the screen which may be explained by figure 6. Rays at angles ϕ_1 and ϕ_2 and Q_1 and R_1 on MM' do not meet unless these are extrapolated behind the screen. It gives a large hatched area beyond the screen resulting into a negative depth of image I. In this case stereoscopic images are combined to form composite 3D image beyond MM' and give rise to a small interaxial angle α. for the same interaxial distance ee' at a given angle of observation θ. This observation reveals that 3D composite image is formed in front of screen if the interaxial angle is large and it goes behind the screen if a is small. This satisfies the stereoscopic condition that α is proportional to perceptive three spatial dimensions of length, breadth and depth.

17

The Physics of Breath-Hold Diving

Some physical features of breath-hold diving are analysed in a simple way. Essentially the diver's descent and the initial surface dive are considered. A few examples are presented which show the diver's buoyancy equilibrium varying with depth, the driving force supplied by finning and the effect of friction between the water and the diver.

The interest in sport-related examples in teaching elementary physics is well know. Very often, a simple physical example taken from sports can be a great help in particle dynamics, fluid mechanics and thermodynamics lectures. This may well be the case for breath - holding diving, also known as snorkelling. It is not something new: long before being considered as a sporting activity, professional breath-hold divers in Korea and Japan . In addition, breath-hold diving is a traditional activity on South Pacific coasts. In the recent decades, the invention of wet suits,and particularly of Neoprene as a manufacturing material have helped make this sport popular and widespread.

From a physical point of view, the following features are relevant in breath-hold diving: the speed of descent and ascent; the driving force needed for both of them; the friction between water and the diver; and the buoyancy equilibrium. All of them have an effect on the time the breath-hold dive will take: two minutes at the most and usually much less for divers without special skills or exceptional physical qualities.

Usual reference to this sport in most elementary physics textbooks deal with buoyancy equilibrium, together with the limitations imposed by hydrostatic pressure to snorkelling . Apparently, no mention is made of all the factors which determine the motion of the breath-hold diver and make it more efficient in terms of saving the divers' energy . We intend here to analyse in a rather simple form, the dynamics of breath-hold diving. A few examples are also presented.

The descent

Let our breath-hold diver have mass *M* and average density ρ_d (defined as the quotient between *M* and the total diver volume outside water). She is supposed to wear a weight belt with mass *m* and density ρ_d. As soon as the she moves down into the water, besides the buoyancy force, a drag force opposes the motion. It can be expressed as $1/2\rho_w C_D Av^2$ where C_D is the density of water, A is the drag coefficient, A is the effective frontal areas of the diver, and V her velocity, If we denote the diver's position by z_1 starting from the surface and taking the downwards direction as positive, the resulting force acting on the diver is

Where V(z) is the total diver's body at depth z. This term deserves some comment. As every diver knows, change in buoyancy with depth (from *positive buoyancy to negative buoyancy,* according to divers'jargon) are due to body compression as an effect of hydrostatic pressure. As the diver's density increases, the upthrust on her body becomes smaller. Every ten metres, pressure increases by approximately one atmosphere. However, the effect of this pressure is not the same on all parts of the body; it is only relevant in the body's air spaces. It is commonly accepted that major changes in body buoyance with depth are due to compression or expansion of air contained in the lungs. Other air spaces such as sinuses, stomach, etc. can hardly change their volume because if they did, rupture and tissue destruction would occur. Therefore,-can be spilt into two contributions, lung air volume-and the volume of the rest of the body, $V_R(z)$

$$V(z) = V_L(z) + V_R(z). \qquad (2)$$

The last term can be taken as constant for the depth reached by breath-hold divers(generally less than 30 m) and be set equal to its surface value:

$$V_R(z) \cong V_R(0) = \frac{M}{\rho_d} - V_L(0). \qquad (3)$$

By assuming that the air in the lungs behaves as an ideal gas, and taking into account that human body temperature remains virtually constant, Boyle's law leads to

$$V_L(z) = \frac{p_0 V_L(0)}{p_0 + \rho_w g z} \qquad (4)$$

Where ρ_o is the atmospheric pressure. Now, let us analyse in detail the very beginning of the divers' motion in order to choose suitable initial conditions. The divers' descent begins in most cases with a move known as *surface dive*. Essentially, it consists in getting into a vertical position from the usual horizontal swimming position. It is accomplished by bending the body at the waist and, at the same time, lifting up the legs so that arms and legs are in a straight line and their weight makes the body to go down into the water. A good surface dive performed by a skillful diver may take her 3-4 m deep without any additional driving force provided by her arms or the fins. The fact that, initially, part of the body is out of the water has not yet been included in the expression for the force. To this purposes, let us consider the diver, as regards her buoyancy as a homogeneous cylinder of length h. According to the position of its centre of mass z, the total submerged volume Vs(z) may be

$$V_s(z) = 0 \quad \text{for } z \leq -\frac{h}{2}$$

(the whole body out of the water) (5a)

$$V_s(z) = \left(\frac{z}{h} + \frac{1}{2}\right)\left(V(z) + \frac{m}{\rho_l}\right) \quad \text{for } -\frac{h}{2} < z < \frac{h}{2}$$

(party under water) (5b)

$$V_s(z) = V(z) + \frac{m}{\rho_l} \quad \text{for } z \geq \frac{h}{2}$$

(the wholw body under water). (5c)

These expressions can be written in an abridged forms as

$$V_s(z) = f(z)\left(V(z) + \frac{m}{\rho_l}\right) \qquad (6)$$

where f(z) is the 'linear step function' defined in equations(5a)-(5c). After substituting for equations (2)-(4) and (6) in the expression in brackets of equation (1), the following resultant force is obtained for the divers' passive descent:

$$\sum F_z = (M+m)g$$
$$- f(z)\left(\frac{p_0 V_L(0)}{p_0+\rho_w g z} + \frac{M}{\rho_d} - V_L(0) + \frac{m}{\rho_l}\right)\rho_w g$$
$$- \frac{\rho_w C_D A}{2} v^2 \qquad (7)$$

Now, if we denote by- the position reached by the divers' centre of mass in the surface dive, we can use as initial condition for solving the above equation- Note that when considering the ascent, the sign of the drag force should be changed.

Some examples

Change in buoyancy. A common experience for any breath-hold diver is breathing out some of the air contained in her lungs and going down at once to the bottom of the swimming pool. This gives us the opportunity of checking how important the body air spaces are as regards buoyancy. From equation, by assuming hydrostatic equilibrium,, we can easily obtain how much air should the diver expel to acquire negative buoyancy. If we assume that the diver wears no weight belt (*m* = 0), together with typical values for ρ_d(970 kg m$_3$), *M*(75kg), ρ_w (1000 kg m^{-3}), and for the air volume in the lungs, we find that approximately 2.3 liters should be breathed out. This quantity increasers with water density (e.g it is slightly greater in sea water, since in this case $\rho_w = 1025$ kg m^{-3} and decreases with the divers' density.

Breath-hold divers usually carry additional weight to make the descent easier. Depending on the depth at which they are going to dive, they put the required amount of weight in their belt so as to have a slightly positive buoyancy at the surface and neutral or slightly negative buoyance at the diving depth. Strictly speaking , in most cases these changes in buoyance come not only from volume change in body air spaces- as we have already seen before - but also from volume changes of the air bubbles contained in wet suits (the main purpose to these bubbles is to slow down the rate of heat loss from the driver's body, since air is a poor conductor of heat). Some breath-hold divers wear a weight belt which can be released after the descent so that they surface more quickly. Too little weight will make the descent difficult, while too much weight will require additional effort to get recovered between successive dives at the surface. The additional mass - needed to get neutral buoyancy at depth z can be readily obtained according to the expression

$$m_n = \frac{\dfrac{p_0 V_L(0)}{p_0 + \rho_w g z} + M\left(\dfrac{1}{\rho_d} - \dfrac{1}{\rho_w}\right) - V_L(0)}{1/\rho_w - 1/\rho_l} \quad (8)$$

Which also comes from equation (7) by taking $F_2 = 0$. Figure shows the dependence of m_n on z for (a) a diver without a wet suit and (b) a diver with a 5 mm thick wet suit(we assume for the diver's density an approximate value fo $P_d = 930$ kg m^{-3}, and the change of wet suit

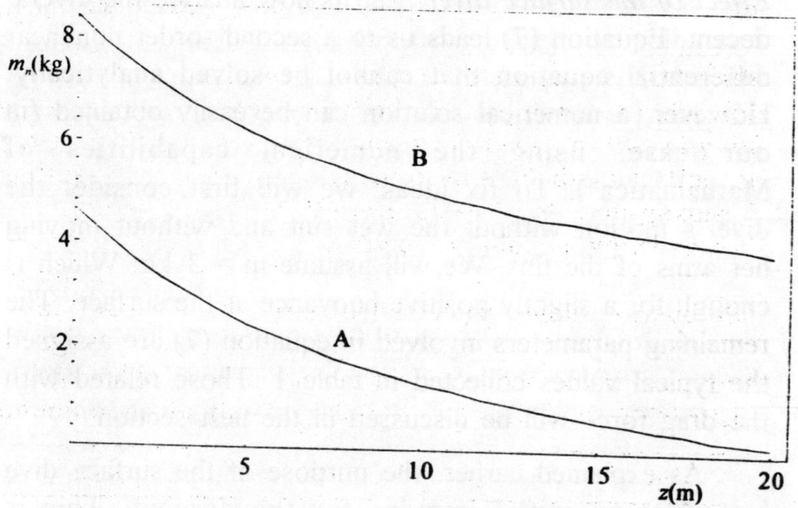

Additional mass needed for a diver of mass 75 kg to get nuetral buoyancy in sea water at depth z

density is neglected). The remaining parameters are given the values shown in table 1.

Table 1. Parameter values used.

ρ_d = 970 kg m^{-3}
ρ_1 = 11340 kg m^{-3}
ρ_w = 1025 kg m^{-3}
ρ_0 = 1.013 x 105 Nm^{-2}
h = 1.75 m
V_L(o) = 6x10-3 m^3
M = 75 kg
C_D = 0.5
A = 0.065 m^2

Effect of the surface diver. Let us now analyse the divers' decent. Equation (7) leads us to a second -order nonlinear differential equation that cannot be solved analytically. However, a numerical solution can be easily obtained (in our case, using the numerical capabilities of Mathematica[R]). To fix ideas, we will first consider the diver's motion without the wet suit and without moving her arms of the fins. We will assume m = 3 kg. Which is enough for a slightly positive buoyance at the surface. The remaining parameters involved in equation (7) are assigned the typical values collected in table 1. Those related with the drag force will be discussed in the nest section.

As explained earlier, the purpose of the surface dive is to get an initial impulse for the descent. This is accomplished by placing the body's centre of mass above its normal equilibrium position in water. The higher the centre of mass is (i.e. the smaller z(0) is) the greater depths she will reach without finning. The function z(t) for several different values of z(0), i.e. for several initial positions of the center of mass, according to the skill of the diver in her surface dive. As we see, in some cases the diver returns to the surface, while in others the initial momentum is large enough to drive her deep. The dotted line shows the equilibrium position for the body's centre of mass. Not that the inflection point of all curves (where is diver's acceleration changes sign) is found at the equilibrium depth.

This 'drop' is something usual of breath-hold divers since it is desirable to reduces as much as possible the body's oxygen consumption associated with mechanical work. It is convenient to save energy in return for some extract time taken in the descent. However, such a drop is not so easy to perform unless the additional weight is accurately chosen. In most cases, divers generally propel themselves with their fins. By taking into account the drag force, a simple calculation shows that the driving force associated with gentle finning (e.g. to ;get a constant velocity around 1 $ms^{-1)}$ should be at least 15 N. The motion of a diver who drives herself downwards; she starts to fin when her buoyancy becomes neutral. In this case a new term *F*(z) is to be added in equation (7) to account for the new driving force.

$$F(z) = 0 \quad \text{for } z < z_n \quad (9)$$

$$F(z) = F_0 \quad \text{for} > z_n$$

Where z_n is the depth of neutral buoyancy, which can be obtained from equation (8). Typical times if descent measured in sea water (see some exprimental data taken by one of the authors) are found for - between 20 N and 40 N.

Effect of the drag force. The effect of the drag force exerted by water is crucial. This is well known by breath-hold divers, who do their best to ensure a good hydrodynamics shape for their body in both the descent and the ascent. Wet suits manufactures also take the dragging force into account, and use preferably Lycra ™ or smooth neoprene for the external cover of wet suits.

The effective frontal area of the diver, A and the drag coefficient C_D, may have a considerable effect on the descent and ascent times. The value A = 0.065 m^2 used in calculations is approximately the surface of an ellipse with semi-axes 10 cm and 20 cm respectively. figure 4 shoes the change in descent trajectory for the first 20 s according to different values of C_D. The experimental

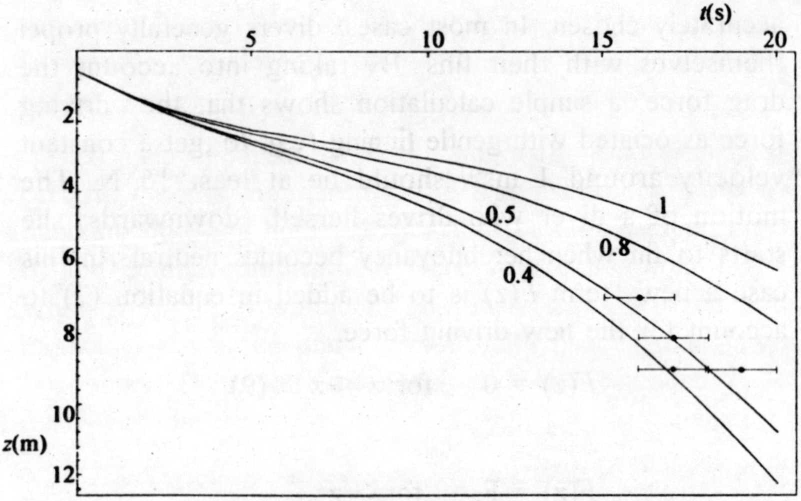

Change in the descent trajectory for the first 20 s, as a function of drag coefficient

measurements shown in the above figure were taken by one of the authors. They seem to support the value of C_D = 0.5 chosen in calculations. Besides, this value for C_D lies with the range found in the literature for bodies of similar shape.

18

Speed of Light Measurement with the Laser Pointer

A recent article in *The Physics Teacher* explained how to modulate the Metrologic laser pointer with an audio signal, a technique that allows the teacher to demonstrate the information-carrying ability of laser light. Teachers can also use a laser pointer to measure the speed of light. Although the idea of measuring the sped of light by incorporating the phase delay between a modulated split beam of light is not new the ability to do it at relatively low cost makes this technique attractive. In this application the laser pointer is modulated at a high frequency. You will need a "travel" distance of about 45 m one way.

On a homemade optical bench the laser is connected to the modulating circuit and turned on. The existing light passes through a beam splitter so 30% of the light detector while the other 70% travels down the hallway and back again. At the point of reflection, a plane mirror mounted on a gimbal mirror mount is aimed so that the reflected light hits a small convex mirror on the optical bench. The second light detector is positioned at the focal point of this lens. By feeding the outputs of these two detector circuits into a dual trace oscilloscope, the two sine wave patterns appear on the screen, but they are out of phase. If the sine wave patterns are centered vertically, the ascending sine waves will cross the x axis at different positions, which correspond to a measurable time. Using this technique, the precise frequency of the modulating circuit is not critical to the success of the experiment. Finally, since the extra distance traveled by the light can be easily measured along the floor with a rape measure, the speed of light can be calculated ($v = d/t$).

When I do the demonstration,the oscilloscope is on the 0.1 μs/cm scale and reads 3.2 cm. The round-trip distance down our hall and back is 87 m. Therefore, the speed of light as calculated is 87 m divided by 0.32 μs = 2.7×10^8 m/s. Circuit diagrams for the oscillator/modulator and detection circuits are shown; a source list is also provided.

19

Measuring Pinhole Images of the Sun

One of the first experiments attempted in many introductory physics laboratories is the "measurement lab". This experiment consist of variations on determining the volume, surface area, or density of assorted solids using rulers, calipers, or micrometers. The primary purpose of this lab is to familiarize students with measurement procedures and to teach the techniques of error analysis that will be used throughout the rest of the course.

One of the great disadvantages of this particular lab is that it confirms most negative student preconceptions of physics. Not only are the actual measurements perceived as rather dull, but additional harassments have been added (in the form of the error analysis); and there is always the suspicion that may be this lab is just killing time until the kinematics lectures can be given, and the real labs assigned.

We would like to propose a more exotic measurement lab-one that will introduce students to a simple phenomenon that they have seen all of their lives, but have never really observed, and one that many will want to show their friends. This experiment will also introduce measurement, and present a simple example of how to use error analysis with an obvious illustration of its value. The particular experiment I describe measures the diameters of pinhole images of the Sum and uses them to calculate the heights of the leafy canopy that created the images.

Description of the Phenomena

When partial eclipses of the Sun are imminent, it is always reported that an easy and safe way to view the eclipse is by making a pinhole and projecting the solar image upon a sheet of paper. Further, if you place a white sheet of paper under a broad-leafed tree, multiple images of the eclipsed Sun will be created by the tapestry of leaves in the canopy above. Some very dramatic pinhole images were observed during the annular eclipse of June 14, 1995 when trees produced multiple images of rings on sidewalks, walls, and ground.

Most people believe that these images can be viewed only during eclipses, and that somehow they are phenomena of the eclipse itself. It is a great revelation when they are shown that Sun images are visible under leafy canopies any time the Sun is visible. (Under special circumstances you can even see large sunspots in the Sun image).

The Experiment

While I leave the physics of the pinhole effect to other sources. The Sun subtends an angle in the sky of about half a degrees. I say "about" because the size of the Sun will depend upon the day of the year (orbital position of

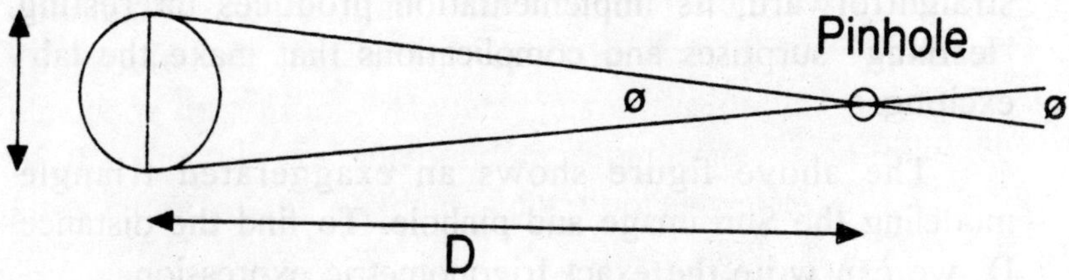

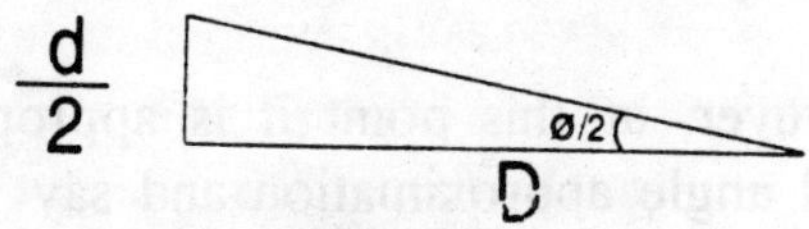

The isosceles triangle created by the image of the sun and the pinhole can be replaced by a right triangles with height d/2 and length D

the Earth). At perihelion (ϕ_p = 0.542 ± 0.0005 deg), the angular size of the Sun will appear larger than at aphelion ((ϕ_a = 0.524 ± 0.0005 deg). This effect can be used to illustrate and example of systematic error, as compared to measurement error (the +0.001 degrees for the angular size of the Sun) associated with all of the measurements.

Because the Sun always shines down at an angle (not directly overhead) in nontropic parts of the world, we see the image of an ellipse upon the ground. If we place a screen where the image appears and tilt it so it directly faces the Sun, the image will become circular. The outline of our experimental procedure will be to measure the diameters *d* of a set of Sun images and use these to calculate the distance *(D)* to the leafy canopy above. Then a direct measurement of *D* will be made using metersticks, and the two values compared to see if they fall within their tolerance limits.

While the experimental procedure is quite

straightforward, its implementation produces interesting "learning" surprises and complications that make the lab exciting.

The above figure shows an exaggerated triangle modeling the Sun image and pinhole. To find the distance D, we can write the exact trigonometric expression

$$D = (d/2)/\tan(\phi/2) \qquad (1)$$

However, at this point it is appropriate to introduce the small angle approximation and say

$$D \approx d/\sin\phi \qquad (2)$$

Realistic numbers will show that the difference between the two calculated values will be far less than the overall error *D*, and thus insignificant.

The way in which ΔD is determined from Eq. (2), for example, will depend on the level of the course and the particular type of error analysis decided for use. However, the measured quantity that is essential in finding ΔD is Δd, which is the error associated with the measurement of the Sun image diameter, Δd. Making this measurement becomes interesting. Unlike the width of a metal block, the boundary of the Sun image will not appear sharp. In fact, the greater the height of the leafy canopy, the "fuzzier" the Sun image will appear. Since there is no "hard" diameter, students are forced into consideration of error analysis. (If the you are in an exceptionally provocative frame of mind, you might even ask your students to make the measurements on a windy day).

The next part of the experiment is to directly measure

ΔD. This is easily done for small trees (using metersticks); the two values of ΔD usually lie within the tolerance of the measured and calculated values. However, all good experiments must have an outstanding opportunity to fail. For large trees, the direct measurement of ΔD must be estimated, so the ΔD value becomes quite large. But more insidiously, you cannot really tell what part of the leafy canopy is forming the pinholes. As a result, even with the larger ΔD error values, the calculated and measured values of ΔD do not always fall within each other's tolerance. This forces students to examine their experimental procedure. Even though the solution to this problem is usually found, it necessitates an explanation in the conclusion section of the student lab report.

As a "first" experiment, the pinhole measurement lab has the positive features of being interesting and novel, of not having major insurmountable difficulties, but at the same time not being trivial: most important, it introduces the concept of error analysis and measurement in such a way as to motivate its necessity.

20

Experiments with "Newton's Cradle"

The toy popularly known as "Newton's Cradle" or "Newton's Balls" consists of a row of five smooth metal balls hanging from pairs of strings attached to a frame. The fascination of the toy comes from its simple dynamics: when the first ball is pulled aside and released, its impact with the row of balls causes one at the other end to fly off, leading to a long-lasting cycle of single balls flying off each end, with the balls in between remaining largely motionless. If we ignore the small amount of damping, the device can serve as a nice illustration of the laws of conservation of momentum and mechanical energy, or even more exotic concepts such as time-reversal invariance.

It is clear that the low damping of this apparatus depends on the high elasticity, or high coefficient of restitution of the balls. What may not be so well appreciated (except by the manufacturer) is how the low damping depends on a precise alignment of the balls. I came to this realization quite by accident when the strings of my set became tangled and one ball remained slightly misaligned with the rest afterwards. In this paper we shall consider the joint effects of elasticity, friction, and ball alignment on the rate of damping of the cycle. The entire exercise could make an excellent lecture demonstration suitable for an overhead projector.

To simplify the analysis, without any great loss of generality, consider an apparatus made from only two balls-or one in which three of the five balls have been pulled out of the way. Further, consider an initial state in which both balls are pulled aside and released from identical angles, so that the lab and center-of-mass systems become identical. In this case, each ball undergoes exactly one half cycle of a full pendulum swing between impacts, so that the frequency of impacts, so that the frequency of impacts (for small angle swings), is predicted to be

$f = g/l \frac{l}{\pi}$ For my particular apparatus, which has a string length $l = 0.12$ m, this yields 173 impacts per minute.

A very convenient way to confirm that the predicated frequency is correct is to activate a metronome at the predicted frequency and observe that the ball collisions occur at a steady frequency exactly matching that of the metronome. Three of the five balls have been pulled aside and rest on a shelf : a cylinder made from a piece of rolled-up index card separates the two hanging balls. The hollow cylinder is held in place by the tangential component of each ball's weight. If the cylinder is quickly snatched away, the balls will collide in a symmetric way. Place the apparatus on an overhead projector with a ruled transparency beneath it, and your class can observe how the amplitude of the maximum ball separation decays with time.

Effect of Inelasticity and Friction on Damping

Suppose that we first consider damping the due only to inelasticity and friction, and assume that the balls collide exactly head on. The coefficient of restitution, may be defined as the ratio of the ball speeds after collision to those before collision in the center of mass system, so that $0 \leq \varepsilon \leq 1$, where the limiting values correspond to completely inelastic and elastic collisions, respectively. If we consider the result of *N* collisions of a pair of equal mass balls, the result is to reduce the initial common speeds of the balls by the factor ε^N– assuming that a constant fraction of the mechanical energy is lost each collision. If no mechanical energy were lost during each ball's swing, we could relating its velocity just after impact to its maximum angle of swing θ

$$V^2 = 2gl\ (1- \cos\theta) \qquad (1)$$

If we assume instead that at the end of each half swing the fraction of the ball's mechanical energy that remains is 8 eq. (1) becomes.

$$V^2\delta = 2gl\ (l\text{-}\cos\theta) \qquad (2)$$

Hence, based on Eq. (2) and the difinitions of ε and δ, we can express the quantity $\varepsilon^2\delta$ in terms of the initial angle of each ball prior to release, θ_o, and their maximum angles after *N* collisions, θ_N:

$$(\varepsilon^2\delta)^N = \frac{v_N^2}{v_0^2} = \frac{1 - \cos\theta_N}{1 - \cos\theta_0} \qquad (3)$$

so that we have

$$\varepsilon^2\delta = \left(\frac{1 - \cos\theta_N}{1 - \cos\theta_0}\right)^{1/N} \qquad (4)$$

Note that $1-\varepsilon^2$ represents the fractional mechanical energy loss during a collision, and $1-\delta$ represents the fractional effects between collisions. Incidentally, there is no theoretical reason to suppose that the fraction or mechanical energy lost during collisions should remain constant over a wide range of velocities - although based on his experiments with the original set of balls, Isaac Newton claimed exactly that. Even through variations in the coefficient of restitution with velocity have been observe, such variations can be relatively small, even over large ranges of velocity. In our experiment, over the relatively limited range we have examined (between V_o and $V_N = 0.707v_o$), the approximation of a constant ε and δ appears to be consistent with observations. The quantity we actually observe as the pair of balls repeatedly collide is not --, But rather the maximum horizontal displacement x of each ball during its swing, which is related to θ by

$$\theta = \sin^{-1}(x/l) \qquad (5)$$

I observed the damping by counting the number of collisions for the initial ball separation (x_o= 1.8 cm) to be reduced to half. This 50- percent amplitude reduction required $N = 158 \pm 10$ collisions, for which application of Eqs. (4) and (5) yields $\varepsilon^2\delta = 0.9912 \pm 0.0008$. Actually, it was a bit too tedious to count up to 158 impacts, so I simply measured the time and found N by using the computed frequency of impacts, which was previously found to be reliable .(based on the metronome measurement noted earlier).

To determine ε and δ separately, simply observe the effects of frictional damping on a freely swinging ball with no collisions ($\varepsilon = 1$), as its horizontal amplitude decays from $x_0 = 1.8$cm to $x_N = 0.9$cm. The number of half swings for this amplitude reduction to occur is found to be $N = 265 \pm 15$, for which Eq. (4) yields $\delta = 0.9969 \pm 0.0001$. Combining this value of δ with the previous value of $\varepsilon^2\delta$ yields $\varepsilon = 0.9969 \pm 0.0004$, meaning that only 100-99.69 = 0.31 percent of the ball's mechanical energy is lost each collision.

Effect of Misalignment of Balls on Damping

As we consider the effects of ball misalignment-again taking the simple case of a two-ball collision in the center of mass frame-the collision is no longer assumed to be head on, and the line joining the ball's centers makes an angle ϕ with each ball's velocity vector. Each velocity initially is assumed to lie along the x-axis, and ϕ is given in terms of the ball's impact parameter d and diameter D by

$$\phi = \sin^{-1} d/D \qquad (6)$$

Immediately after the collision, each ball recoils along directions making an angle 2ϕ with the positive or negative x-axis. If we ignored inelasticity and friction, the ball's speed immediately before and after impact would be the same. However, because the balls are suspended by pairs of strings, they are free to swing only along a specific direction that we have defined as the x-direction. Thus, a short time after the collision only the x-component of each ball's velocity ($v\cos 2\phi$) remains, because the constraining strings quickly reduce the y-component ($v \sin 2\phi$) to zero. The net effect is to reduce the ratio of the ball speeds shortly before and after a collision by the factor $\cos 2\phi$. After N collisions, the reduction factor is

When friction (during the swing) and inelasticity (during the impact) are accounted for, the ratio instead becomes

$$\frac{v_N}{v_0} = (\varepsilon\delta^{1/2} \cos 2\phi)^N \qquad (8)$$

If we want the damping due to ball misalignment to be less than that due to the combined effect of inelasticity and friction, we require that

$$\cos 2\phi > \varepsilon\delta^{1/2}$$

which with the aid of Eq. (6) yields

$$d/D < 2 \sin\left(\frac{1}{2} \cos^{-1} \varepsilon\delta^{1/2}\right) \qquad (10)$$

Inequality (10) tells us how much one ball misaligned by (as a fraction of its diameter) for damping due to misalignment to be less than that due to misalignment to be less than that due to the unavoidable inelasticity and friction. In a sense, it represents the alignment a designer

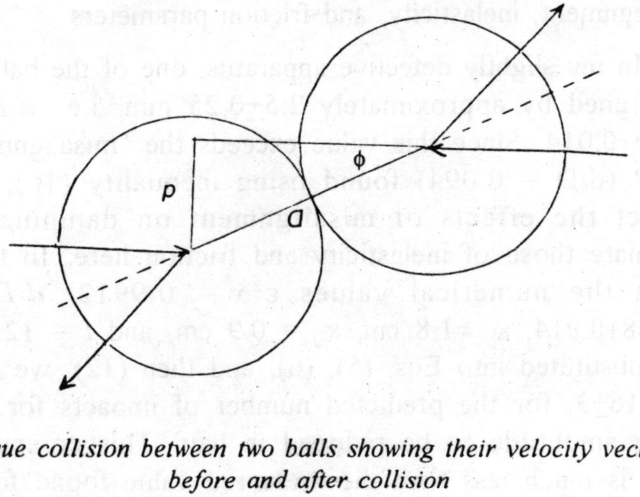

Oblique collision between two balls showing their velocity vectors before and after collision

of this fascinating toy should strive for. We can apply inequality (10) to obtain a misalignment limit, given the measured value of $\varepsilon^2\delta = 0.9912$ and ball diameters of D = 18 mm. We find $d/D < 0.094$, or therefore d<1.7mm.

How much damping is created due to a specific combination of misalignment, inelasticity, and friction? By combining Eqs. (4) and (8) we find that

$$(\varepsilon^2\delta)^N \cos^{2N} 2\phi = \frac{v_N^2}{v_0^2} = \frac{1 - \cos\theta_N}{1 - \cos\theta_0} \quad (11)$$

We can then easily solve Eq. (11) for N, yielding

$$N = \frac{\ln(1 - \cos\theta_N) - \ln(1 - \cos\theta_0)}{\ln(\varepsilon^2\delta\cos^2 2\phi)} \quad (12)$$

with ϕ given by Eq. (6), and θ and θ_N given by Eq. (5). The value of N given in Eq. (12) represents the predicted number of collisions required for the amplitude to be reduced from θ to θ_N expressed in terms of values of the misalignment, inelasticity, and friction parameters.

In my slightly defective apparatus, one of the balls is misaligned by approximately 2.5$\pm$0.25 mm, i.e., d/D = 0.139$\pm$0.014. Since this value exceeds the "misalignment limit" (d/D = 0.094) found using inequality (10), we expect the effects of misalignment on damping to dominate those of inelasticity and friction here. In fact, when the numerical values $\varepsilon^2\delta = 0.9912$, d/D = 0.1388+0.014, x_0 =1.8 cm, x_N = 0.9 cm, and l = 12 cm are substituted into Eqs. (5), (6), and then (12), we find N = 16$\pm$3, for the predicted number of impacts for the swing amplitude to be reduced in half. This computed value is much less than the measured value found for a pair of aligned balls (N = 158), and it is in excellent agreement with the measured value for the colliding misaligned balls (N = 20 $\pm$2). The close agreement with theory verifies our model that it is only the x-component of momentum that is conserved in the process.

Index